The IEE Wiring Regulations
Explained and Illustrated

By the same author

**Wiring Systems and Fault Finding
for Installation Electricians**

Electrical Installation Work

The IEE Wiring Regulations Explained and Illustrated

Second Edition

Brian Scaddan, IEng, MIElecIE

 NEWNES

To my mother

Newnes
An imprint of Butterworth-Heinemann Ltd
Linacre House, Jordan Hill, Oxford OX2 8DP

 PART OF REED INTERNATIONAL BOOKS

OXFORD   LONDON   BOSTON
MUNICH   NEW DELHI   SINGAPORE   SYDNEY
TOKYO   TORONTO   WELLINGTON

First published 1989
Reprinted 1989, 1990 (twice)
Second edition 1991
Reprinted 1992

British Library Cataloguing in Publication Data
Scaddan, Brian
  The IEE wiring regulations
  1. Buildings. Electric equipment.
  Installation. Regulations: Institute of
  Electrical Engineers. Regulations for
  electrical installations – Critical studies
  I. Title
  621.319′24

ISBN 0 7506 0314 3

Printed in England by Clays Ltd, St Ives plc

Contents

Preface

During the past seven years, in developing and teaching courses devoted to compliance with the IEE Wiring Regulations, it has become apparent to me that many operatives and personnel in the electrical contracting industry have forgotten the basic principles and concepts upon which electric power supply and its use are based. As a result of this, misconceived ideas and much confusion have arisen over the interpretation of the Regulations.

It is the intention in this book to dispel such misconceptions and to educate and where necessary refresh the memory of the reader. In this respect, emphasis has been placed on those areas where most confusion arises, namely earthing and bonding, protection, and circuit design. Much of Part 5 of the Regulations is not mentioned, since it deals with selection of accessories etc. which needs little or no explanation. The same applies to Part 7 which deals with special situations.

The sixteenth edition of the IEE Wiring Regulations, to which this book now conforms, was published in June 1991, although the fifteenth edition is to run in parallel until it is finally superseded in January 1993. The philosophy and concepts that the book seeks to explain remain unchanged, regardless of the edition. It is *not* a guide to the Regulations or a replacement for them; nor does it seek to interpret them Regulation by Regulation. It should, in fact, be read in conjunction with them; to help the reader, each chapter cites the relevant Regulation numbers for cross-reference.

It is hoped that the book will be found particularly useful by college students, electricians and technicians, and also by managers of smaller electrical contracting firms that do not normally employ engineers or designers.

Introduction: the IEE Wiring Regulations

It was once said, by whom I have no idea, that 'rules and regulations are for the guidance of wise men and the blind obedience of fools.' This is certainly true in the case of the IEE Regulations for the Electrical Equipment of Buildings – commonly known as the IEE Wiring Regulations. They are *not* statutory rules, but recommendations for the safe selection and erection of wiring installations. Earlier editions were treated as an 'electrician's Bible': the Regulations now take the form primarily of a design document.

The IEE Wiring Regulations are divided into seven parts. These follow a logical pattern from the basic requirements to the final testing and inspection of an installation:

Part 1 indicates the range and type of installations covered by the Regulations, what they are intended for, and the basic requirements for safety.

Part 2 is devoted to the definitions of the terms used throughout the Regulations.

Part 3 details the general information needed before any design work can usefully proceed.

Part 4 informs the designer of the different methods available for protection against electric shock, overcurrent etc., and how to apply those methods.

Part 5 enables the correct type of equipment, cable, accessory etc. to be selected in accordance with the requirements of Parts 1 to 4.

Part 6 deals with particular requirements for special installations such as bathrooms, swimming pools, construction sites etc.

Part 7 provides details of the relevant tests to be performed on a completed installation before it is energized.

Appendices 1 *to* 6 provide tabulated and other background information required by the designer.

It must be remembered that the Regulations are not a collection of unrelated statements each to be interpreted in isolation; there are many cross-references throughout which may render such an interpretation valueless.

In using the Regulations I have found the index an invaluable starting place when seeking information. However, one may have to try different combinations of wording in order to locate a particular item. For example, determining how often an RCD should be tested via its test button could prove difficult since no reference is made under 'Residual current devices' or 'Testing'; however, 'Periodic testing' leads to Regulation 514–12, and the information in question is found in 514–12–02. In the index, this Regulation is referred to under 'Notices'.

ONE

Fundamental requirements for safety

[IEE Regs Part 1 and Chapter 13]

It does not require a degree in electrical engineering to realize that electricity at *low* voltage can, if uncontrolled, present a serious threat of injury to persons or livestock, or damage to property by fire.

Clearly the type and arrangement of the equipment used, together with the quality of workmanship provided, will go a long way to minimizing danger. The following is a list of basic requirements:

1 Use good workmanship.

2 Use approved materials and equipment.

3 Ensure that the correct type, size and current-carrying capacity of cables is chosen.

4 Ensure that equipment is suitable for the maximum power demanded of it.

5 Make sure that conductors are insulated, and sheathed or protected if necessary, or are placed in a position to prevent danger.

6 Joints and connections should be properly constructed to be mechanically and electrically sound.

7 Always provide overcurrent protection for every circuit in an installation (the protection for the whole installation is usually provided by the supply authority), and ensure that protective

devices are suitably chosen for their location and the duty they have to perform.

8 Where there is a chance of metalwork becoming live owing to a fault, it should be earthed, and the circuit concerned should be protected by an overcurrent device or a residual current device (RCD).

9 Ensure that all necessary bonding of services is carried out.

10 Do not place a fuse, a switch or a circuit breaker, unless it is a linked switch or circuit breaker, in an earthed neutral conductor. The linked type must be arranged to break all the phase conductors.

11 All single-pole switches must be wired in the phase conductor only.

12 A readily accessible and effective means of isolation must be provided, so that all voltage may be cut off from an installation or any of its circuits.

13 All motors must have a readily accessible means of disconnection.

14 Ensure that any item of equipment which may normally need operating or attending to by persons is accessible and easily operated.

15 Any equipment required to be installed in a situation exposed to weather or corrosion, or in explosive or volatile environments, should be of the correct type for such adverse conditions.

16 Before adding to or altering an installation, ensure that such work will not impair any part of the existing installation.

17 After completion of an installation or an alteration to an installation, the work must be inspected and tested to ensure, as far as reasonably practicable, that the fundamental requirements for safety have been met.

These requirements form the basis of the IEE Regulations.

It is interesting to note that, whilst the Wiring Regulations are not statutory, they are based on a Statutory document: the Electricity Supply Regulations 1988. Apart from all these requirements, it must also be remembered that we are bound by the Electricity at Work Regulations, and the Health and Safety at Work Act, both of which of course are Statutory. In fact, the Health and Safety Executive produces guidance notes for installations in such places as schools and construction sites. The contents of these documents reinforce and extend the requirements of the IEE Regulations. Extracts from the Health and Safety at Work Act and the Electricity at Work Regulations are reproduced below.

The Health and Safety at Work Act 1974

Duties of employers

Employers must safeguard, as far as is reasonably practicable, the health, safety and welfare of all the people who work for them. This applies in particular to the provision and maintenance of safe plant and systems of work, and covers all machinery, equipment and appliances used.

Some examples of the matters which many employers need to consider are:

1 Is all plant up to the necessary standards with repect to safety and risk to health?

2 When new plant is installed, is latest good practice taken into account?

3 Are systems of work safe? Thorough checks of all operations, especially those operations carried out infrequently, will ensure that danger of injury or to health is minimized. This may require special safety systems, such as 'permits to work'.

4 Is the work environment regularly monitored to ensure that, where known toxic contaminants are present, protection conforms to current hygiene standards?

5   Is monitoring also carried out to check the adequacy of control measures?

6   Is safety equipment regularly inspected? All equipment and appliances for safety and health, such as personal protective equipment, dust and fume extraction, guards, safe access arrangement, monitoring and testing devices, need regular inspection (Section 2(1) and 2(2) of the Act).

No charge may be levied on any employee for anything done or provided to meet any specific requirement for health and safety at work (Section 9).

Risks to health from the use, storage, or transport of 'articles' and 'substances' must be minimized. The term *substance* is defined as 'any natural or artificial substance whether in solid or liquid form or in the form of gas or vapour' (Section 53(1)).

To meet these aims, all reasonably practicable precautions must be taken in the handling of any substance likely to cause a risk to health. Expert advice can be sought on the correct labelling of substances, and the suitability of containers and handling devices. All storage and transport arrangements should be kept under review.

Safety information and training

It is now the duty of employers to provide any necessary information and training in safe practices, including information on legal requirements.

Duties to others

Employers must also have regard for the health and safety of the self-employed or contractors' employees who may be working close to their own employees; and for the health and safety of the public who may be affected by their firm's activities.

Similar responsibilities apply to self-employed persons, manufacturers and suppliers.

4

Duties of employees

Employees have a duty under the Act to take reasonable care to avoid injury to themselves or to others by their work activities, and to cooperate with employers and others in meeting statutory requirements. The Act also requires employees not to interfere with or misuse anything provided to protect their health, safety or welfare in compliance with the Act.

Electricity at Work Regulations

Persons on whom duties are imposed by these Regulations

(1) Except where otherwise expressly provided in these Regulations, it shall be the duty of every:

  (a) employer and self-employed person to comply with the provisions of these Regulations in so far as they relate to matters which are within his control; and

  (b) manager of a mine or quarry (within in either case the meaning of section 180 of the Mines and Quarries Act 1954[a]) to ensure that all requirements or prohibitions imposed by or under these Regulations are complied with in so far as they relate to the mine or quarry or part of a quarry of which he is the manager and to matters which are within his control.

(2) It shall be the duty of every employee while at work:

  (a) to cooperate with his employer so far as is necessary to enable any duty placed on that employer by the provisions of these Regulations to be complied with; and

  (b) to comply with the provisions of these Regulations in so far as they relate to matters which are within his control.

---

[a] 1954 C.70; section 180 was amended by S1 1974/2013.

Employer

1 For the purposes of the Regulations, an employer is any person or body who (a) employs one or more individuals under a contract of employment or apprenticeship; or (b) provides training under the schemes to which the HSW Act applies through the Health and Safety (Training for Employment) Regulations 1988 (Statutory Instrument No 1988/1222).

Self-employed

2 A self-employed person is an individual who works for gain or reward otherwise than under a contract of employment whether or not he employs others.

Employee

3 Regulation 3(2)(a) reiterates the duty placed on employees by section 7(b) of the HSW Act.

4 Regulation 3(2)(b) places duties on employees equivalent to those placed on employers and self-employed persons where these are matters within their control. This will include those trainees who will be considered as employees under the Regulations described in paragraph 1.

5 This arrangement recognises the level of responsibility which many employees in the electrical trades and professions are expected to take on as part of their job. The 'control' which they exercise over the electrical safety in any particular circumstances will determine to what extent they hold responsibilities under the Regulations to ensure that the Regulations are complied with.

6 A person may find himself responsible for causing danger to arise elsewhere in an electrical system, at a point beyond his own installation. This situation may arise, for example, due to unauthorized or unscheduled back feeding from his installation onto the system, or to raising the fault power level on the system above rated and agreed maximum levels due to connecting extra

generation capacity, etc. Because such circumstances are 'within his control', the effect of regulation 3 is to bring responsibilities for compliance with the rest of the regulations to that person, thus making him a duty holer.

Absolute/reasonably practicable

7 Duties in some of the regulations are subject to the qualifying term 'reasonably practicable'. Where qualifying terms are absent the requirement in the regulation is said to be absolute. The meaning of reasonably practicable has been well established in law. The interpretations below are given only as a guide to duty holders.

Absolute

8 If the requirement in a regulation is 'absolute', for example if the requirement is not qualified by the words 'so far as is reasonably practicable', the requirement must be met regardless of cost or any other consideration. Certain of the regulations making such absolute requirements are subject to the Defence provision of regulation 29.

Reasonably practicable

9 Someone who is required to do something 'so far as is reasonably practicable' must assess, on the one hand, the magnitude of the risks of a particular work activity or environment and, on the other hand, the costs in terms of the physical difficulty, time, trouble and expense which would be involved in taking steps to eliminate or minimize those risks. If, for example, the risks to health and safety of a particular work process are very low, and the cost or technical difficulties of taking certain steps to prevent those risks are very high, it might not be reasonably practicable to take those steps. The greater the degree of risk, the less weight that can be given to the cost of measures needed to prevent that risk.

10 In the context of the Regulations, where the risk is very often that of death, for example, from electrocution, and where the nature of the precautions which can be taken are so often very simple and cheap, e.g. insulation, the level of duty to prevent that danger approaches that of an absolute duty.

11 The comparison does not include the financial standing of the duty holder. Furthermore, where someone is prosecuted for failing to comply with a duty 'so far as is reasonably practicable', it would be for the accused to show the court that it was not reasonably practicable for him to do more than he had in fact done to comply with the duty (section 40 of the HSW Act).

TWO

Earthing

[Relevant chapters and parts
Chapters 31, 33, 41, 47, 53, 54, 55, Part 6]

Definitions used in this chapter

Bonding conductor   A protective conductor providing equipotential bonding.

Circuit protective conductor   A protective conductor connecting exposed conductive parts of equipment to the main earthing terminal.

Direct contact   Contact of persons or livestock with live parts which may result in electric shock.

Earth   The conductive mass of earth, whose electric potential at any point is conventionally taken as zero.

Earth electrode resistance   The resistance of an earth electrode to earth.

Earth fault loop impedance   The impedance of the phase-to-earth loop path starting and ending at the point of fault.

Earthing conductor   A protective conductor connecting a main earthing terminal of an installation to an earth electrode or other means of earthing.

Earth leakage current   A current which flows to earth, or to extraneous conductive parts, in a circuit which is electrically sound.

Equipotential bonding   Electrical connection putting various exposed conductive parts and extraneous conductive parts at a substantially equal potential.

Exposed conductive part   A conductive part of equipment which can be touched and which is not a live part but which may become live under fault conditions.

Extraneous conductive part   A conductive part liable to introduce a potential, generally earth potential and not forming part of the electrical installation.

Functional earthing   Connection to earth necessary for proper functioning of electrical equipment.

Indirect contact   Contact of persons or livestock with exposed conductive parts made live by a fault and which may result in electric shock.

Live part   A conductor or conductive part intended to be energized in normal use, including a neutral conductor but, by convention, not a PEN conductor.

PEN conductor   A conductor combining the functions of both protective conductor and neutral conductor.

Phase conductor   A conductor of an AC system for the transmission of electrical energy, other than a neutral conductor.

Protective conductor   A conductor used for some measure of protection against electric shock and intended for connecting together any of the following parts:

   exposed conductive parts
   extraneous conductive parts
   Main earthing terminal
   Earth electrode(s)
   Earthed point of the source.

Residual current device   A mechanical switching device or association of devices intended to cause the opening of the contacts when the residual current attains a given value under given conditions.

Simultaneously accessibe parts   Conductors or conductive parts which can be touched simultaneously by a person or, where applicable, by livestock.

Earth: what it is, and why and how we connect to it
[IEE Regs section 413]

The thin layer of material which covers our planet, be it rock, clay, chalk or whatever, is what we in the world of electricity refer to as earth. So, why do we need to connect anything to it? After all, it is not as if earth is a good conductor.

Perhaps it would be wise at this stage to investigate potential difference (PD). A potential difference is exactly what it says it is: a difference in potential (volts). Hence two conductors having PDs of, say, 20 V and 26 V have a PD between them of $26 - 20 = 6$ V. The original PDs, i.e. 20 V and 26 V, are the PDs between 20 V and 0 V and 26 V and 0 V.

So where does this 0 V or zero potential come from? The simple answer is, in our case, the earth. The definition of earth is therefore the conductive mass of earth, whose electric potential at any point is conventionally taken as zero.

Hence if we connect a voltmeter between a live part (e.g. the phase conductor of, say, a socket outlet) and earth, we would read 240 V; the conductor is at 240 V, the earth at zero. Of course it must be remembered that we are discussing the supply industry in the UK, where earth potential is very important. We would measure nothing at all if we connected our voltmeter between, say, the positive 12 V terminal of a car battery and earth, as in this case the earth plays no part in any circuit. Figure 1 illustrates this difference.

Note the connection of the supply neutral in Figure 1a to earth, which makes it possible to have a complete circuit via the earth. Supply authority neutrals should be at around zero volts, and in

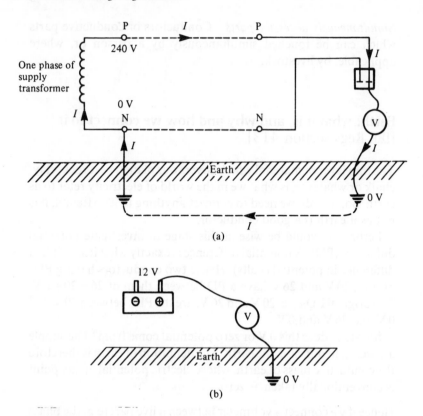

Figure 1

order to maintain this condition they are connected to the zero potential of earth.

This also means that a person in an installation touching a live part whilst standing on the earth would take the place of the voltmeter in Figure 1a, and could suffer a severe electric shock. Remember that the accepted *lethal* level of shock current passing through a person is only 50 mA or 1/20 A. The same situation would arise if the person were touching, say, a faulty appliance and a gas or water pipe (Figure 2).

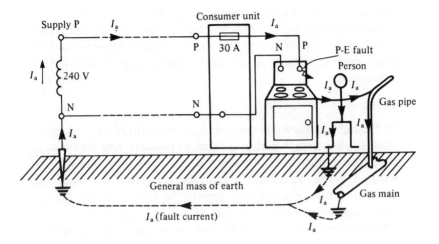

Figure 2

One method of providing some measure of protection against these effects is to join together (bond) all metallic parts and connect them to earth. This ensures that all metalwork in a healthy situation is at or near zero volts, and under fault conditions all metalwork will rise to the same potential. So, simultaneous contact with two such metal parts would not result in a shock, as there will be no PD between them. This method is known as earthed equipotential bonding.

Unfortunately, as previously mentioned, earth itself is not a good conductor unless it is very wet, and therefore it presents a high resistance to the flow of fault current. This resistance is usually enough to restrict fault current to a level well below that of the rating of the protective device, leaving a faulty circuit uninterrupted. Clearly this is an unhealthy situation. The methods of overcoming this problem will be dealt with later.

In all but the most rural areas, consumers can connect to a metallic earth return conductor which is ultimately connected to the earthed neutral of the supply. This, of course, presents a low-resistance path for fault currents to operate the protection.

Summarizing, then, connecting metalwork to earth places that metal at or near zero potential, and bonding between metallic parts puts such parts at the same potential even under fault conditions.

Connecting to earth

In the light of previous comments, it is obviously necessary to have as low an earth path resistance as possible, and the point of connection to earth is one place where such resistance may be reduced. When two conducting surfaces are placed in contact with each other, there will be a resistance to the flow of current dependent on the surface areas in contact. It is clear, then, that the greater surface contact area with earth that can be achieved, the better.

There are several methods of making a connection to earth, including the use of rods, plates and tapes. By far the most popular method in everyday use is the rod earth electrode. The plate type needs to be buried at a sufficient depth to be effective and, as such plates may be 1 or 2 metres square, considerable excavation may be necessary. The tape type is predominantly used in the earthing of large electricity substations, where the tape is laid in trenches in a mesh formation over the whole site. Items of plant are then earthed to this mesh.

Rod electrodes

These are usually of solid copper or copper-clad carbon steel, the latter being used for the larger-diameter rods with extension facilities. These facilities comprise: a thread at each end of the rod to enable a coupler to be used for connection of the next rod; a steel cap to protect the thread from damage when the rod is being driven in; a steel driving tip; and a clamp for the connection of an earth tape or conductor (Figure 3).

The choice of length and diameter of such a rod will, as previously mentioned, depend on the soil conditions. For example, a long thick electrode is used for earth with little moisure retention.

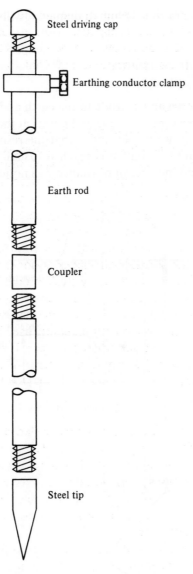

Steel driving cap

Earthing conductor clamp

Earth rod

Coupler

Steel tip

Figure 3

Generally, a 1–2 m rod, 16 mm in diameter, will give a relatively low resistance.

Earth electrode resistance
[IEE Regs Definitions]

If we were to place an electrode in the earth and to measure the resistance between the electrode and points at increasingly larger distances from it, we would notice that the resistance increased with distance until a point was reached (usually around 2.5 m) beyond which no increase in resistance was noticed (Figure 4).

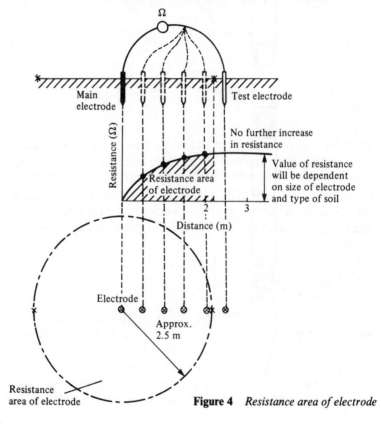

Figure 4   *Resistance area of electrode*

The resistance area around the electrode is particularly important with regard to the voltage at the surface of the ground (Figure 5). For a 2 m rod, with its top at ground level, 80–90 per cent of the voltage appearing at the electrode under fault conditions is

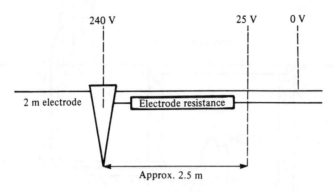

240 V   25 V   0 V

2 m electrode   Electrode resistance

Approx. 2.5 m

Figure 5

dropped across the earth in the first 2.5 m to 3 m. This is particularly dangerous where livestock is present, as the hind and fore legs of an animal can be respectively inside and outside the resistance area: a PD of 25 V can be lethal! One method of overcoming this problem is to house the electrode in a pit below ground level (Figure 6) as this prevents voltages appearing at ground level.

Earthing in the IEE Regulations
[IEE Regs Section 547]
[IEE Regs 413–02–01 to 28]

In the preceding pages we have briefly discussed the reasons for, and the importance and methods of, earthing. Let us now examine the subject in relation to the IEE Regulations.

Contact with metalwork made live by a fault is called *indirect contact*. One popular method of providing some measure of protection against such contact is by earthed equipotential bonding

17

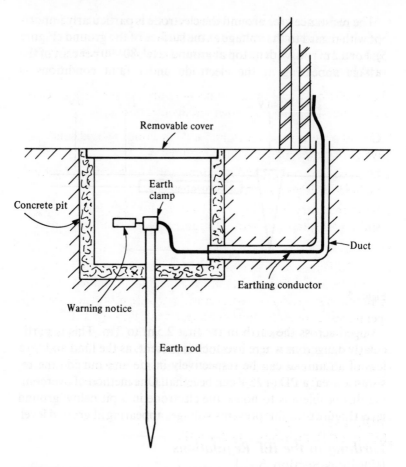

Figure 6

and automatic disconnection of supply. This entails the bonding together and connection to earth of:

1 All metalwork associated with electrical apparatus and systems, termed exposed conductive parts. Examples include conduit, trunking and the metal cases of apparatus.

2 All metalwork liable to introduce a potential including earth potential, termed extraneous conductive parts. Examples are gas, oil and water pipes, structural steelwork, radiators, sinks and baths.

The conductors used in such connections are called *protective conductors*, and they can be further subdivided into:

1 Circuit protective conductors, for connecting exposed conductive parts to the main earthing terminal;
2 Main equipotential bonding conductors, for bonding together main incoming services, structural steelwork etc.; and
3 Supplementary bonding conductors, for bonding together sinks, baths, taps, radiators etc. and exposed conductive parts.

The effect of all this bonding is to create a zone in which all metalwork of different services and systems will, even under fault conditions, be at a substantially equal potential. If, added to this, there is a low-resistance earth return path, the protection should operate fast enough to prevent danger. [IEE Reg 413–02–04].

The resistance of such an earth return path will depend upon the system (see the next section), either TT, TN–S or TN–C–S (IT systems will not be discussed here, as they are extremely rare and unlikely to be encountered by the average contractor).

Earthing systems
[IEE Regs Definitions (Systems)]

These have been designated in the IEE Regulations using the letters T, N, C and S. These letters stand for:

T   terre (French for earth) and meaning a direct connection to earth
N   neutral
C   combined
S   separate

When these letters are grouped they form the classification of a type of system. The first letter in such a classification denotes how the supply source is earthed. The second denotes how the metalwork of an installation is earthed. The third and fourth indicate the functions of neutral and protective conductors. Hence:

1 A TT system has a direct connection of the supply source to earth and a direct connection of the installation metalwork to earth. An example is an overhead line supply with earth electrodes, and the mass of earth as a return path (Figure 7).

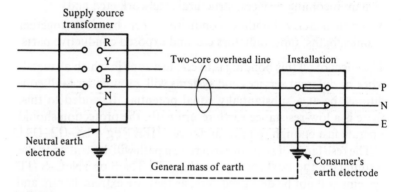

Figure 7 *TT system*

2 A TN–S system has the supply source directly connected to earth, the installation metalwork connected to the earthed neutral of the supply source via the lead sheath of the supply cable, and the neutral and protective conductors throughout the whole system performing separate functions (Figure 8).

3 A TN–C–S system is as the TN–S but the supply cable sheath is also the neutral, i.e. it forms a combined earth/neutral conductor known as a PEN (protective earthed neutral) conductor (Figure 9). The installation earth and neutral are separate conductors. This system is also known as PME (protective multiple earthing).

20

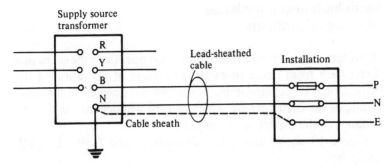

Figure 8 *TN–S* system

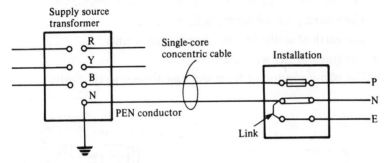

Figure 9 *TN–C–S system*

Note that only single-phase systems have been shown, for simplicity.

Summary

In order to avoid the risk of serious electric shock, it is important to provide a path for earth leakage currents to operate the circuit protection, and to endeavour to maintain all metalwork at the same potential. This is achieved by bonding together metalwork of electrical and non-electrical systems to earth. The path for leakage currents would then be via the earth itself in TT systems or by a metallic return path in TN–S or TN–C–S systems.

21

Earth fault loop impedance
[IEE Regs Definitions]

As we have seen, circuit protection should operate in the event of a direct fault from phase to earth. The speed of operation of the protection is of extreme importance and will depend on the magnitude of the fault current, which in turn will depend on the impedance of the earth fault loop path.

Figure 10 shows this path. Starting at the fault, the path comprises:

1 The circuit protective conductor (CPC).
2 The consumer's earthing terminal and earth conductor.
3 The return path, either metallic or earth.
4 The earthed neutral of the supply transformer.
5 The transformer winding.
6 The phase conductor from the transformer to the fault.

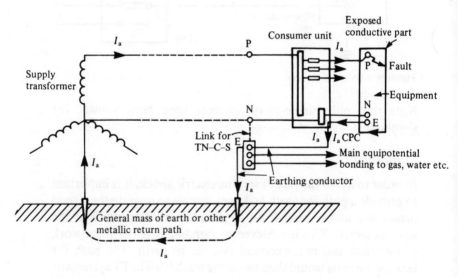

Figure 10

Figure 11 is a simplified version of this path. We have

$$Z_s = Z_e + R_1 + R_2$$

where Z_s is the actual total loop impendance, Z_e is the loop impedance external to the installation, R_1 is the resistance of the phase conductor, and R_2 is the resistance of the CPC. We also have

$$I_a = U_0/Z_s$$

where I_a is the fault current and U_0 is the nominal voltage to earth (usually 240 V).

Determining the value of total loop impedance
[IEE Regs 413–02–04]

The IEE Regulations require that when the general characteristics of an installation are assessed, the loop impedance Z_e external to the installation shall be ascertained.

This may be measured in existing installations using a phase-to-earth loop impedance tester. However, when a building is only at the drawing board stage it is clearly impossible to make such a measurement. In this case, we have three methods available to assess the value of Z_e:

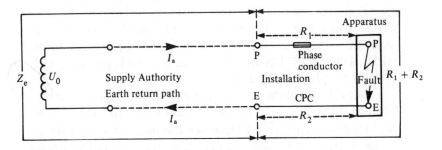

Figure 11

23

1 Determine it from details (if available) of the supply transformer, the main distribution cable and the proposed service cable; or
2 Measure it from the supply intake position of an adjacent building having service cable of similar size and length to that proposed; or
3 Use maximum likely values issued by the supply authority as follows:
TT system: 21 ohms maximum
TN–S system: 0.8 ohms maximum
TN–C–S system: 0.35 ohms maximum.

Method 1 will be difficult for anyone except engineers. Method 3 can, in some cases, result in pessimistically large cable sizes. Method 2, if it is possible to be used, will give a closer and more realistic estimation of Z_e. However, if in any doubt, use method 3.

Having established a value for Z_e, it is now necessary to determine the impedance of that part of the loop path internal to the installation. This is, as we have seen, the resistance of the phase conductor plus the resistance of the CPC, i.e. $R_1 + R_2$. Resistances of copper conductors may be found from manufacturers' information which gives values of resistance/metre for copper and aluminium conductors at 20°C in milliohms/metre.

It should be noted that a copper conductor has a set resistance, no matter by what name it is called. Hence the phase conductor figures given without a CPC size, e.g. 16.00, will also be the value for a 16 mm CPC. This enables us to find $R_1 + R_2$ for non-standard arrangements. For example, a 25 mm phase conductor with a 4 mm^2 CPC has $R_1 = 0.727$ and $R_2 = 4.61$, giving $R_1 + R_2$ $= 0.727 + 4.61 = 5.337$ mΩ/m.

So, having established a value for $R_1 + R_2$, we must now multiply it by the length of the run and divide by 1000 (the values given are in milliohms per metre). However, this final value is based on a temperature of 20°C; under fault conditions, when large currents flow, the conductor temperature will increase. In order to get an approximate value of resistance at the higher temperature, a multiplier is used.

Hence, for a 20 m length of PVC insulated 16 mm² phase conductor with a 4 mm² CPC, the value of $R_1 + R_2$ would be

$$R_1 + R_2 = (1.15 + 4.61) \times 20 \times 1.38/1000 = 0.159 \text{ ohms}$$

The factor 1.38 is the multiplier for PVC insulation.

We are now in a position to determine the total earth fault loop impedance Z_s from

$$Z_s = Z_e + R_1 + R_2$$

As previously mentioned, this value of Z_s should be as low as possible to allow enough fault current to flow to operate the protection as quickly as possible. Tables 41B1, B2 and D of the IEE Regulations give maximum values of loop impedance for different sizes and types of protection for both socket outlet circuits and bathrooms, and circuits feeding fixed equipment.

Provided that the actual values calculated do not exceed those tabulated, socket outlet circuits will disconnect under earth fault conditions in 0.4 s or less, and circuits feeding fixed equipment in 5 s or less. The reasoning behind these different times is based on the time that a faulty circuit can reasonably be left uninterrupted. Hence socket outlet circuits from which hand-held appliances may be used, and bathrooms with their high water content, clearly present a greater shock risk than circuits feeding fixed equipment.

It should be noted that these times, i.e. 0.4 s and 5 s, do not indicate the duration that a person can be in contact with a fault. They are based on the probable chances of someone being in contact with exposed or extraneous conductive parts at the precise moment that a fault develops.

See also Tables 41A, 604A and 605A of the IEE Regulations.

Example 1

Let us now have a look at a typical example of, say, a shower circuit run in an 18 m length of 6.0 mm² (6242Y) twin cable with CPC, and protected by a 30 A BS 3036 semi-enclosed rewirable fuse. A 6.0 mm² twin cable has a 2.5 mm² CPC. We will also assume that the external loop impedance Z_e is measured as 0.27 ohms. Will there be a shock risk if a phase-to-earth fault occurs?

The total loop impedance $Z_s = Z_e + R_1 + R_2$. We are given $Z_e = 0.27$ ohms. For a 6.0 mm² phase conductor with a 2.5 mm² CPC, $R_1 + R_2$ is 10.49 mΩ/m. Hence, with a multiplier of 1.38 for PVC,

$$\text{total } R_1 + R_2 = 18 \times 10.49 \times 1.38/1000 = 0.26 \text{ ohms}$$

Therefore, $Z_s = 0.27 + 0.26 = 0.53$ ohms. This is less than the 1.14 ohms maximum given in Table 41B1 for a 30 A BS 3036 fuse. Hence the protection will disconnect the circuit in less than 0.4 s. In fact it will disconnect in less than 0.1 s, but the determination of this time will be dealt with in Chapter 5.

Example 2

Consider now, a more complex installation, and note how the procedure remains unchanged.

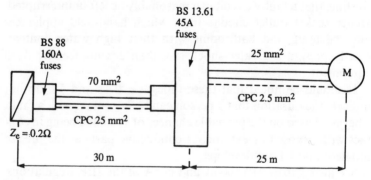

Figure 12

In this example, a 3-phase motor is fed using $25\,\text{mm}^2$ single PVC conductors in trunking, the CPC being $2.5\,\text{mm}^2$. The circuit is protected by BS 1361 45A fuses in a distribution fuseboard. The distribution circuit or sub-main feeding this fuseboard comprises $70\,\text{mm}^2$ PVC singles in trunking with a $25\,\text{mm}^2$ CPC the protection being by BS 88 160A fuses. The external loop impedance Z_e has a measured value of 0.2Ω. Will this circuit arrangement comply with the shock risk constraints?

The formula $Z_s = Z_e + R_1 + R_2$ must be extended, as the $(R_1 + R_2)$ component comprises both sub-main and motor circuit, it therefore becomes:

$$Z_s = Z_e + (R_1 + R_2)_1 + (R_1 + R_2)_2$$

Distribution circuit $(R_1 + R_2)_1$

This comprises 30 m of $70\,\text{mm}^2$ single-phase conductor and 30 m of $25\,\text{mm}^2$ CPC. Typical values for conductors over $35\,\text{mm}^2$ are shown in the following table:

Area of conductor (mm^2)	Resistance in mΩ/m Copper	Aluminium
50	0.387	0.641
70	0.263	0.443
95	0.193	0.320
120	0.153	0.253
150	0.124	0.206
185	0.0991	0.164
240	0.0754	0.125
300	0.0601	0.1

As an alternative we can use our knowledge of the relationship between conductor resistance and area, e.g. a $10\,\text{mm}^2$ conductor has approximately 10 times less resistance than a $1\,\text{mm}^2$ conductor:

$10\,\text{mm}^2$ resistance $= 1.83\,\text{m}\Omega/\text{m}$
$1\,\text{mm}^2$ resistance $= 18.1\text{m}\Omega/\text{m}$

Hence a $70\,\text{mm}^2$ conductor will have a resistance approximately half that of a $35\,\text{mm}^2$ conductor.

$35\,\text{mm}^2$ resistance $= 0.524\,\text{m}\Omega/\text{m}$

$\therefore\ 70\,\text{mm}^2$ resistance $= \dfrac{0.524}{2} = 0.262\,\text{m}\Omega/\text{m}$

which compares well with the value given in the table.

$25\,\text{mm}^2$ CPC resistance $= 0.727\,\text{m}\Omega/\text{m}$

so the distribution circuit

$$(R_1 + R_2) = 30 \times (0.262 + 0.727) \times 1.38/1000 = 0.041\,\Omega.$$

Hence $Z_s = Z_e + (R_1 + R_2)_1 = 0.2 + 0.041 = 0.241\,\Omega$ which is less than the Z_s maximum of $0.267\,\Omega$ quoted for a 160A BS 88 fuse in Table 41D.

Motor circuit $(R_1 + R_2)_2$

Here we have $25\,\text{m}$ of $25\,\text{mm}^2$ conductor with $25\,\text{m}$ of $2.5\,\text{mm}^2$ CPC. Hence:

$$(R_1 + R_2)_2 = 25 \times (0.727 + 7.41) \times 1.38/1000$$
$$= 0.28\,\Omega$$
$$\therefore\ \text{Total } Z_s = Z_e + (R_1 + R_2)_1 + (R_1 + R_2)_2$$
$$= 0.2 + 0.041 + 0.28$$
$$= 0.521\,\Omega$$

which is less than the Z_s maximum of $1.0\,\Omega$ quoted for a 45A BS 1361 fuse from Table 41D.

Hence we have achieved compliance with the shock-risk constraints.

Residual current devices
[IEE Regs 542–11] [IEE Regs 412–06, 413–02–15 and 471–16–0]

We have seen how very important is the total earth loop impedance Z_s in the reduction of shock risk. However, in TT systems where the mass of earth is part of the fault path, the maximum values of Z_s given in Tables 41B1, B2 and D may be hard to satisfy. Added to this, climatic conditions will alter the resistance of the earth in such a way that Z_s may be satisfactory in wet weather but not in very dry.

The IEE Regulations recommend therefore that protection for socket outlet circuits in a TT system be achieved by a residual current device (RCD), such that the product of its residual operating current and the loop impedance will not exceed a figure of 50 V. Residual current breakers (RCBs), residual current circuit breakers (RCCBs) and RCDs are one and the same thing.

For construction sites and agricultural environments this value is reduced to 25 V.

Principle of operation of an RCD

Figure 13 illustrates the construction of an RCD. In a healthy circuit the same current passes through the phase coil, the load,

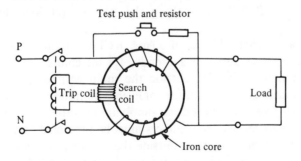

Figure 13   *Residual current device*

and back through the neutral coil. Hence the magnetic effects of phase and neutral currents cancel out.

In a faulty circuit, either phase to earth or neutral to earth, these currents are no longer equal. Therefore the out-of-balance current produces some residual magnetism in the core. As this magnetism is alternating, it links with the turns of the search coil, inducing an EMF in it. This EMF in turn drives a current through the trip coil, causing operation of the tripping mechanism.

It should be noted that a phase-to-neutral fault will appear as a load, and hence the RCD will not operate for this fault.

Nuisance tripping

Certain appliances such as cookers, water heaters and freezers tend to have, by the nature of their construction and use, some leakage currents to earth. These are quite normal, but could cause the operation of an RCD protecting an entire installation. This can be overcome by using split-load consumer units, where socket outlet circuits are protected by a 30 mA RCD, leaving all other circuits controlled by a normal mains switch. Better still, especially in TT systems, is the use of a 100 mA RCD for protecting circuits other than socket outlets.

Modern developments in MCB, RCD and consumer unit design now make it easy to protect any individual circuit with a combined MCB/RCD, making the use of split-load boards unnecessary.

One area where the use of 30 mA RCDs is required is in the protection of socket outlets intended for the connection of portable appliances for use outside the main equipotential zone. Hence socket outlets in garages or even within the main premises which are likely to be used for supplying portable tools such as lawn mowers and hedge trimmers must be protected by a 30 mA RCD. All other equipment outside the main equipotential zone should, in the event of an earth fault, disconnect in 0.4 s.

An exception to the RCD requirement is where fixed equipment is connected to the supply via a socket outlet, provided that some means of preventing the socket outlet being used for hand-held appliances is ensured.

Supplementary bonding
[IEE Regs Section 547–03]

This is perhaps the most debated topic in the IEE Regulations. The confusion may have arisen because of a lack of understanding of earthing and bonding. Hopefully, this chapter will rectify the situation.

By now we should know why bonding is necessary; the next question however, is to what extent bonding should be carried out. This is perhaps answered best by means of question and answer examples:

1 Q: Why do I need to bond the hot and cold taps and a metal sink together? Surely they are all joined anyway?
   A: In most sinks the holes for connection of the taps are usually surrounded by a plastic insert which tends to insulate the taps from the sink. The sink is a large metallic part which is very 'earthy', especially when water is running from the sink outlet. The hot and cold taps are both parts of different systems, but are once again earthy owing not least to the water in them. These therefore are all extraneous conductive parts and should be bonded together.

2 Q: Do I have to bond all the radiators in a premises with a protective conductor run back to the main earth terminal?
   A: Supplementary bonding is only necessary when extraneous conductive parts are simultaneously accessible with other extraneous conductive parts and where they are not already connected to the main equipotential bonding by a metallic path of a permanent and reliable nature, which can be an extraneous conductive part. So, even if two radiators are simultaneously ac-

cessible with each other or with, say, a metal-clad socket outlet, providing that the pipework feeding them is metal and has soldered connections and the socket outlet is correctly connected to a CPC, there is no need to bond.

3 Q: Do I need to bond metal window frames?
A: In general, no. Apart from the fact that most window frames will not introduce a potential from anywhere, the part of the window most likely to be touched is the opening portion, to which it would not be practicable to bond. There may be a case for the bonding of patio doors, which could be considered earthy with rain running from the lower portion to the earth. However, once again the part most likely to be touched is the sliding section, to which it is not possible to bond. In any case there would need to be another simultaneously accessible part to warrant considering any bonding.

4 Q: What about bonding in bathrooms?
A: Bathrooms are particularly hazardous areas with regard to shock risk, as body resistance is drastically reduced when wet. Hence bonding between simultaneously accessible exposed conductive parts must be carried out in addition to their existing CPCs. Also of course, taps and metal baths need bonding together, and to other simultaneously accessible extraneous and exposed conductive parts. It may be of interest to note that in older premises a toilet basin may be connected into a cast iron collar which then tees outside into a cast iron soil pipe. This arrangement will clearly introduce earth potential into the bathroom, and hence the collar should be bonded to any simultaneously accessible conductive parts. This may require an unsightly copper earth strap.

5 Q: What size of bonding conductors should I use?
A: Main equipotential bonding conductors should be not less than half the size of the main earthing conductor, subject to a minimum of $6.0 \, \text{mm}^2$ or, where PME (TNCS) conditions are

present, 10.0 mm². For example, most new domestic installations now have a 16.00 mm² earthing conductor, so all main
bonding will be in 10.0 mm². Supplementary bonding conductors are subject to a minimum of 2.5 mm² if mechanically
protected or 4.0 mm² if not. However, if these bonding conductors are connected to exposed conductive parts, they must be
the same size as the CPC connected to the exposed conductive
part, once again subject to the minimum sizes mentioned. It is
sometimes difficult to protect a bonding conductor mechanically throughout its length, and especially at terminations, so it
is perhaps better to use 4.0 mm² as the minimum size.

6 Q: Do I have to bond free-standing metal cabinets, screens,
workbenches etc.?
A: No. These items will not introduce a potential into the equipotential zone from outside, and cannot therefore be regarded
as extraneous conductive parts.

The Faraday cage

In one of his many experiments, Michael Faraday (1791–1867)
placed himself in an open-sided cube which was then covered in a
conducting material and insulated from the floor. When this cage
arrangement was charged to a high voltage, he found that he
could move freely within it touching any of the sides, with no
adverse effects. He had in fact created an equipotential zone, and
of course in a correctly bonded installation we live and/or work in
Faraday cages!

Problems

1 What is the resistance of a 10 m length of 6.0 mm² copper phase
conductor if the associated CPC is 1.5 mm²?

2 What is the length of a 6.0 mm² copper phase conductor with a
2.5 mm² CPC if the overall resistance is 0.261 ohms?

3 If the total loop impedance of a circuit is 0.96 ohms and the cable is a 20 m length of 4.0 mm^2 copper with a 1.5 mm^2 CPC, what is the external loop impedance?

4 Will there be a shock risk if a double socket outlet, fed by a 23 m length of 2.5 mm^2 copper conductor with a 1.5 mm^2 CPC, is protected by a 20 A BS 3036 rewirable fuse and the external loop impedance is measured as 0.5 ohms?

5 A cooker control unit incorporating a socket outlet is protected by a 30 A BS 3871 type 2 MCB, and wired in 6.0 mm^2 copper with a 2.5 mm^2 CPC. The run is some 25 m and the external loop impedance of the TNS system is not known. Is there a shock risk, and if so how could it be rectified?

THREE

Protection

[Relevant IEE parts, chapters and sections
Part 4; Chapters 41, 42, 43, 45. Section 471, 473,
514, 522, 543, 547]

Definitions used in this chapter

Arm's reach   A zone of accessibility to touch, extending from any point on a surface where persons usually stand or move about, to the limits which a person can reach with his hand in any direction without assistance.

Barrier   A part providing a defined degree of protection against contact with live parts, from any usual direction.

Class 2 equipment   Equipment in which protection against electric shock does not rely on basic insulation only, but in which additional safety precautions such as supplementary insulation are provided. There is no provision for the connection of exposed metalwork of the equipment to a protective conductor, and no reliance upon precautions to be taken in the fixed wiring of the installation.

Circuit protective conductor   A protective conductor connecting exposed conductive parts of equipment to the main earthing terminal.

Design current   The magnitude of the current intended to be carried by a circuit in normal service.

Direct contact   Contact of persons or livestock with live parts which may result in electric shock.

Enclosure   A part providing an appropriate degree of protection of equipment against certain external influences and a defined degree of protection against contact with live parts from any direction.

Exposed conductive part   A conductive part of equipment which can be touched and which is not a live part but which may become live under fault conditions.

Extraneous conductive part   A conductive part liable to introduce a potential, generally earth potential, and not forming part of the electrical installation.

Fault current   A current resulting from a fault.

Fixed equipment   Equipment fastened to a support or otherwise secured in a specific location.

Indirect contact   Contact of persons or livestock with exposed conductive parts made live by a fault and which may result in electric shock.

Isolation   Cutting off an electrical installation, a circuit or an item of equipment from every source of electrical energy.

Insulation   Suitable non-conductive material enclosing, surrounding or supporting a conductor.

Live part   A conductor or conductive part intended to be energized in normal use, including a neutral conductor but, by convention, not a PEN conductor.

Obstacle   A part preventing unintentional contact with live parts but not preventing deliberate contact.

Overcurrent   A current exceeding the rated value. For conductors the rated value is the current-carrying capacity.

Overload   An overcurrent occurring in a circuit which is electrically sound.

Residual current device   A mechanical switching device or association of devices intended to cause the opening of the contacts when the residual current attains a given value under specified conditions.

Short-circuit current An overcurrent resulting from a fault of negligible impedance between live conductors having a difference of potential under normal operating conditions.

Skilled person A person with technical knowledge or sufficient experience to enable him to avoid the dangers which electricity may create.

What is protection?

The meaning of the word 'protection', as used in the electrical industry, is no different to that in everyday use. People protect themselves against personal or financial loss by means of insurance and from injury or discomfort by the use of the correct protective clothing. They further protect their property by the installation of security measures such as locks and/or alarm systems. In the same way, electrical systems need

1 To be protected against mechanical damage, the effects of the environment and electrical overcurrents; and

2 To be installed in such a fashion that persons and/or livestock are protected from the dangers that such an electrical installation may create.

Let us now look at these protective measures in more detail.

Protection against mechanical damage

The word 'mechanical' is somewhat misleading in that most of us associate it with machinery of some sort. In fact a serious electrical overcurrent left uninterrupted for too long can cause distortion of conductors and degradation of insulation; both of these effects are considered to be mechanical damage.

However, let us start by considering the ways of preventing mechanical damage by physical impact and the like.

Cable construction

A cable comprises one or more conductors each covered with an insulating material. This insulation provides protection from shock by 'direct contact and prevents the passage of leakage currents between conductors.

Clearly, insulation is very important and in itself should be protected from damage. This may be achieved by covering the insulated conductors with a protective sheathing during manufacture, or by enclosing them in conduit or trunking at the installation stage.

The type of sheathing chosen and/or the installation method will depend on the environment in which the cable is to be installed. For example, metal conduit with PVC singles or mineral insulated (MI) cable would be used in preference to PVC sheathed cable clipped direct, in an industrial environment. Figure 14 shows the effect of physical impact on MI cable.

Protection against corrosion

Mechanical damage to cable sheaths and metalwork of wiring systems can occur through corrosion, and hence care must be taken to choose corrosion-resistant materials and to avoid contact between dissimilar metals in damp situations.

Figure 14   *Mineral-insulated cable. On impact, all parts including the conductors are flattened, and a proportionate thickness of insulation remains between conductors, and conductors and sheath, without impairing the performance of the cable at normal working voltages*

Protection against thermal effects

This is the subject of Chapter 42 of the IEE Regulations. It basically requires common-sense decisions regarding the placing of fixed equipment, such that surrounding materials are not at risk from damage by heat.

Added to these requirements is the need to protect persons from burns by guarding parts of equipment liable to exceed temperatures listed in Table 42A.

Polyvinyl chloride

Polyvinyl chloride (PVC) is a thermoplastic polymer widely used in electrical installation work for cable insulation, conduit and trunking. General-purpose PVC is manufactured to the British Standard BS 6746.

PVC in its raw state is a white powder; it is only after the addition of plasticizers and stabilizers that it acquires the form that we are familiar with.

Degradation

All PVC polymers are degraded or reduced in quality by heat and light. Special stablilizers added during manufacture help to retard this degradation at high temperatures. However, it is recommended in the IEE Regulations that PVC-sheathed cables or thermoplastic fittings for luminaires (light fittings) should not be installed where the temperature is likely to rise above 60°C. Cables insulated with high-temperature PVC (up to 80°C) should be used for drops to lampholders and entries into batten-holders. PVC conduit and trunking should not be used in temperatures above 60°C.

Embrittlement and cracking

PVC exposed to low temperatures becomes brittle and will easily crack if stressed. Although both rigid and flexible PVC used in cables and conduit can reach as low as − 5°C without becoming

39

brittle, the Regulations recommend that general-purpose PVC-insulated cables should not be installed in areas where the temperature is likely to be consistently below 0°C. They further recommend that PVC-insulated cable should not be handled unless the ambient temperature is above 0°C and unless the cable temperature has been above 0°C for at least 24 hours.

Where rigid PVC conduit is to be installed in areas where the ambient temperature is below − 5°C but not lower than − 25°C, type B conduit manufactured to BS 4607 should be used.

When PVC-insulated cables are installed in loft spaces insulated with polystyrene granules, contact between the two polymers can cause the plasticizer in the PVC to migrate to the granules. This causes the PVC to harden and although there is no change in the electrical properties, the insulation may crack if disturbed.

Protection against ingress of solid objects and liquid

In order to protect equipment from damage by foreign bodies or liquid, and also to prevent persons from coming into contact with live or moving parts, such equipment is housed inside an enclosure.

The degree of protection offered by such an enclosure is indicated by an index of protection (IP) code, as shown in the accompanying table. It will be seen from this table that, for instance, an enclosure to IP56 is dustproof and waterproof.

IP codes

First numeral: mechanical protection

0   No protection of persons against contact with live or moving parts inside the enclosure. No protection of equipment against ingress of solid foreign bodies.

1   Protection against accidental or inadvertent contact with live or moving parts inside the enclosure by a large surface of the human body, for example a hand, but not protection against deliberate access to such parts. Protection against ingress of large solid foreign bodies.

2   Protection against contact with live or moving parts inside the enclosure by fingers. Protection against ingress of medium-size solid foreign bodies.

3   Protection against contact with live or moving parts inside the enclosures by tools, wires or such objects of thickness greater than 2.5 mm. Protection against ingress of small foreign bodies.

4   Protection against contact with live or moving parts inside the enclosure by tools, wires or such objects of thickness greater than 1 mm. Protection against ingress of small solid foreign bodies.

5   Complete protection against contact with live or moving parts inside the enclosure. Protection against harmful deposits of dust. The ingress of dust is not totally prevented, but dust cannot enter in an amount sufficient to interfere with satisfactory operation of the equipment enclosed.

6   Complete protection against contact with live or moving parts inside the enclosures. Protection against ingress of dust.

Second numeral: liquid protection

0   No protection.

1   Protection against drops of condensed water. Drops of condensed water falling on the enclosure shall have no harmful effect.

2   Protection against drops of liquid. Drops of falling liquid shall have no harmful effect when the enclosure is tilted at any angle up to 15° from the vertical.

3   Protection against rain. Water falling in rain at an angle equal to or smaller than 60° with respect to the vertical shall have no harmful effect.

4   Protection against splashing. Liquid splashed from any direction shall have no harmful effect.

5   Protection against water jets. Water projected by a nozzle from any direction under stated conditions shall have no harmful effect.

6   Protection against conditions on ships' decks (deck with watertight equipment). Water from heavy seas shall not enter the enclosures under prescribed conditions.

7   Protection against immersion in water. It must not be possible for water to enter the enclosure under stated conditions of pressure and time.

8   Protection against indefinite immersion in water under specified pressure. It must not be possible for water to enter the enclosure.

X   Indicates no *specified* protection.

IEE Regulations 522–01 to 31 give details of the types of equipment, cables, enclosures etc. that may be selected for certain environmental conditions.

Protection against electric shock
[IEE Regs. Chapter 41 and Section 471]

There are two ways of receiving an electric shock: by direct contact, and by indirect contact. It is obvious that we need to provide protection against both of these conditions.

Protection against direct contact
[IEE Regs Section 412, and Regs 471–04 to 08]

Clearly, it is not satisfactory to have live parts accessible to touch by persons or livestock. The IEE Regulations recommend five ways of minimizing this danger:

1 By covering the live part or parts with insulation which can only be removed by destruction, e.g. cable insulation.
2 By placing the live part or parts behind a barrier or inside an enclosure providing protection to at least IP2X. In most cases, during the life of an installation it becomes necessary to open an enclosure or remove a barrier. Under these circumstances, this action should only be possible by the use of a key or tool, e.g. by using a screwdriver to open a junction box. Alternatively access should only be gained after the supply to the live parts has been disconnected, e.g. by isolation on the front of a control panel where the cover cannot be removed until the isolator is in the 'off' position. An intermediate barrier of at least IP2X will give protection when an enclosure is opened: a good example of this is the barrier inside distribution fuseboards, preventing accidental contact with incoming live feeds.
3 By placing obstacles to prevent unintentional approach to or contact with live parts. This method must only be used where skilled persons are working.
4 By placing out of arm's reach: for example, the high level of the bare conductors of travelling cranes.

5 By using an RCD. Whilst not permitted as the sole means of protection, this is considered to reduce the risk associated with direct contact, provided that one of the other methods just mentioned is applied, and that the RCD has a rated operating current of not more than 30 mA and an operating time not exceeding 40 ms at 150 mA.

Protection against indirect contact
[IEE Regs Section 413 and Regs 471–09 to 12]

The IEE Regulations suggest five ways of protecting against indirect contact. One of these – earthed equipotential bonding and automatic disconnection of supply – has already been discussed in Chapter 2. The other methods are as follows.

Use of class 2 equipment

Often referred to as double-insulated equipment, this is typical of modern DIY tools where there is no provision for the connection of a CPC. This does not mean that there should be no exposed conductive parts and that the casing of equipment should be of an insulating material; it simply indicates that live parts are so well insulated that faults from live to conductive parts cannot occur.

Non-conducting location

This is basically an area in which the floor, walls and ceiling are all insulated. Within such an area there must be no protective conductors, and socket outlets will have no earthing connections.

It must not be possible simultaneously to touch two exposed conductive parts, or an exposed conductive part and an extraneous conductive part. This requirement clearly prevents shock current passing through a person in the event of an earth fault, and the insulated construction prevents shock current passing to earth.

Earth-free local equipotential bonding

This is in essence a Faraday cage, where all metal is bonded together but *not* to earth. Obviously great care must be taken when entering such a zone in order to avoid differences in potential between inside and outside.

The areas mentioned in this and the previous method are very uncommon. Where they do exist, they should be under constant supervision to ensure that no additions or alterations can lessen the protection intended.

Electrical separation

This method relies on a supply from a safety source such as an isolating transformer to BS 3535 which has no earth connection on the secondary side. In the event of a circuit that is supplied from such a source developing a live fault to an exposed conductive part, there would be no path for shock current to flow: see Figure 15.

Once again, great care must be taken to maintain the integrity of this type of system, as an inadvertent connection to earth, or interconnection with other circuits, would render the protection useless.

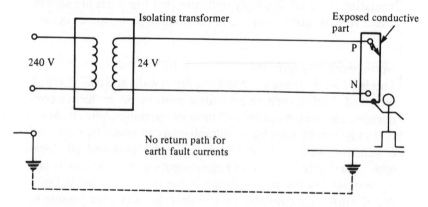

Figure 15

Exemptions
[IEE Regs 471–13]

As with most sets of rules and regulations, there are certain areas which are exempt from the requirements. These are listed quite clearly in IEE Regulations 471–13, and there is no point in repeating them all here. However, one example is the dispensing of the need to earth exposed conductive parts such as small fixings, screws and rivets, provided that they cannot be touched or gripped by a major part of the human body (not less than 50 mm by 50 mm), and that it is difficult to make and maintain an earth connection.

Protection against direct and indirect contact
[IEE Regs Section 411]

So far we have dealt separately with direct and indirect contact. However, we can protect against both of these conditions with the following methods.

Safety extra low voltage (SELV)
This is simply extra low voltage (less than 50 V AC) derived from a safety source such as a class 2 safety isolating transformer to BS 3535; or a motor generator which has the same degree of isolation as the transformer; or a battery or diesel generator; or an electronic device such as a signal generator.

Live or exposed conductive parts of SELV circuits should not be connected to earth, or protective conductors of other circuits, and SELV circuit conductors should ideally be kept separate from those of other circuits. If this is not possible, then the SELV conductors should be insulated to the highest voltage present.

Obviously, plugs and sockets of SELV circuits should not be interchangeable with those of other circuits.

SELV circuits supplying socket outlets are mainly used for hand lamps or soldering irons, for example in schools and colleges. Perhaps a more common example of a SELV circuit is a domestic bell installation, where the transformer is to BS 3535. Note that bell wire is usually only suitable for 50–60 V, which means that it should *not* be run together with circuit cables of higher voltages.

Reduced voltage systems
[IEE Regs 471–15]

The Health and Safety Executive accepts that a voltage of 65 V to earth, three phase, or 55 V to earth, single phase, will give protection against severe electric shock. They therefore recommend that portable tools used on construction sites etc. be fed from a 110 V centre-tapped transformer to BS 4343. Figure 16 shows how 55 V is derived. Earth fault loop impedance values for these systems may be taken from Table 471A.

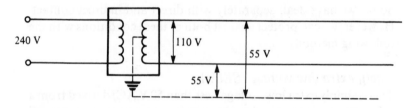

Figure 16

Protection against overcurrent
[IEE Regs Chap 43 and Definitions]

An overcurrent is a current greater than the rated current of a circuit. It may occur in two ways:

1 As an overload current; or
2 As a short-circuit or fault current.

These conditions need to be protected against in order to avoid damage to circuit conductors and equipment. In practice, fuses and circuit breakers will fulfil both of these needs.

Overloads

Overloads are overcurrents occurring in healthy circuits. They may be caused, for example, by faulty appliances or by surges due to motors starting.

Short circuits

A short-circuit current is the current that will flow when a 'dead short' occurs between live conductors (phase to neutral for single phase; phase to phase for three phase). Prospective short-circuit current is the same, but the term is usually used to signify the value of short-circuit current at fuse or circuit breaker positions.

Prospective short-circuit current is of great importance. However, before discussing it or any other overcurrent further, it is perhaps wise to refresh our memories with regard to fuses and circuit breakers and their characteristics.

Fuses and circuit breakers

As we all know, a fuse is the weak link in a circuit which will break when too much current flows, thus protecting the circuit conductors from damage.

There are many different types and sizes of fuse, all designed to perform a certain function. The IEE Regulations refer to only four of these: BS 3036, BS 88, BS 1361 and BS 1362 fuses. It is perhaps sensible to include, at this point, circuit breakers to BS 3871.

Breaking capacity of fuses and circuit breakers
[IEE Reg 130–03–01]

When a short circuit occurs, the current may, for a fraction of a second, reach hundreds or even thousands of amperes. The protective device must be able to break or make such a current without damage to its surroundings by arcing, overheating or the scattering of hot particles.

47

British Standards for fuse links

Standard	Current rating	Voltage rating
BS 2950	Range 0.05 to 25 A	Range 1000 V (0.05 A) to 32 V (25 A) AC and DC
BS 646	1, 2, 3 and 5 A	Up to 250 V AC and DC
BS 1362 cartridge	1, 2, 3, 5, 7, 10 and 13 A	Up to 250 V AC
BS 1361 HRC cut-out fuses	5, 15, 20, 30 and 45 A 60 A	Up to 250 V AC
BS 88 motors	Four ranges, 2 to 1200 A	Up to 660 B, but normally 250 or 415 V AC and 250 or 500 V DC
BS 2692	Main range from 5 to 200 A; 0.5 to 3 A for voltage transformer protective fuses	Range from 2.2 kV to 132 kV
BS 3036 rewirable	5, 15, 20, 30, 45, 60, 100, 150 and 200 A	Up to 250 V to earth
BS 4265	500 mA to 6.3 A 32 mA to 2 A	Up to 250 V AC

British Standards for fuse links (*continued*)

Breaking capacity	Notes
Two or three times current rating	Cartridge fuse links for telecommunication and light electrical apparatus. Very low breaking capacity
1000 A	Cartridge fuse intended for fused plugs and adapters to BS546: 'round-pin' plugs
6000 A	Cartridge fuse primarily intended for BS1363: 'flat pin' plugs
16 500 A 33 000 A	Cartridge fuse intended for use in domestic consumer units. The dimensions prevent interchangeability of fuse links which are not of the same current rating
Ranges from 10 000 to 80 000 A in four AC and three DC categories	Part 1 of Standard gives performance and dimensions of cartridge fuse links, whilst Part 2 gives performance and requirements of fuse carriers and fuse bases designed to accommodate fuse links complying with Part 1
Ranges from 25 to 750 MVA (main range) 50 to 2500 MVA (VT fuses)	Fuses for AC power circuits above 660 V
Ranges from 1000 to 12 000 A	Semi-enclosed fuses (the element is a replacement wire) for AC and DC circuits
1500 A (high breaking capacity) 35 A (low breaking capacity)	Miniature fuse links for protection of appliances of up to 250 V (metric Standard)

Special standard not included above:
AU 105: Fuse (35 A and 50 A) for vehicle electrical equipment
G 176: Cartridge fuses for aircraft
BS 714: Cartridge fuse links for use in railway signalling circuits

The accompanying table indicates the performance of the more commonly used British Standard fuse links.

Fuse and circuit breaker operation
[IEE Regs 433–02 and 434–03]

Let us consider a protective device rated at, say, 10 A. This value of current can be carried indefinitely by the device, and is known as its nominal setting I_n. The value of the current which will cause operation of the device, I_2, will be larger than I_n, and will be dependent on the device's *fusing factor*. This is a figure which, when multiplied by the nominal setting I_n, will indicate the value of operating current I_2.

For fuses to BS 88 and BS 1361 and circuit breakers to BS 3871, this fusing factor is approximately 1.45; hence our 10 A device would not operate until the current reached $1.45 \times 10 = 14.5$ A. The IEE Regulations require co-ordination between conductors and protection when an overload occurs, such that:

1 The nominal setting of the device I_n is greater than or equal to the design current of the circuit I_b ($I_n \geqslant I_b$).
2 The nominal setting I_n is less than or equal to the lowest current-carrying capacity I_z, of any of the circuit conductors ($I_n \leqslant I_z$).
3 The operating current of the device, I_2 is less than or equal to $1.45 I_z$ ($I_2 \leqslant 1.45 I_z$).

So, for our 10 A device, if the cable is rated at 10 A then condition 2 is satisfied. Since the fusing factor is 1.45, condition 3 is also satisfied: $I_2 = I_n \times 1.45 = 10 \times 1.45$, which is also 1.45 times the 10 A cable rating.

The problem arises when a BS 3036 semi-enclosed rewirable fuse is used, as it may have a fusing factor of as much as 2. In order to comply with condition 3, I_n should be less than or equal to $0.725 I_z$. This figure is derived from $1.45/2 = 0.725$. For example, if a cable is rated at 10 A, then I_n for a BS 3036 should be $\leqslant 0.725 \times 10 = 7.25$ A. As the fusing factor is 2, the operating current $I_2 = 2 \times 7.25 = 14.5$, which conforms with condition 3, i.e. $I_2 \leqslant 1.45 \times 120 = 14.5$.

All of these foregoing requirements ensure that conductor insulation is undamaged when an overload occurs.

Under short-circuit conditions it is the conductor itself that is susceptible to damage and must be protected. Figure 17 shows one half-cycle of short-circuit current if there were no protection. The RMS value (0.7071 × maximum value) is called the prospective short-circuit current. The cut-off point is where the short-circuit current is interrupted and an arc is formed; the time t_1 taken to reach this point is called the pre-arcing time. After the current has been cut off, it falls to zero as the arc is being extinguished. The time t_2 is the total time taken to disconnect the fault.

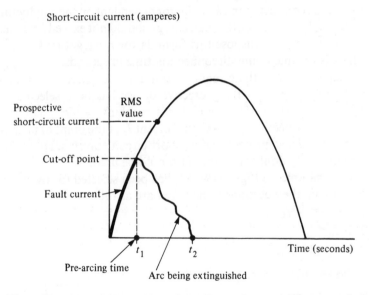

Figure 17

During the time t_1, the protective device is allowing energy to pass through to the load side of the circuit. This energy is known as the pre-arcing let-through energy and is given by $I_f^2 t_1$, where I_f is the short-circuit current. The total let-through energy from start to disconnection of the fault is given by $I_f^2 t_2$ (see Figure 18).

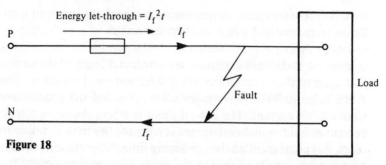

Figure 18

For faults of up to 5 s duration, the amount of heat energy that a cable can withstand is given by k^2s^2, where s is the cross-sectional area of the conductor and k is a factor dependent on the conductor material. Hence the let-through energy should not exceed k^2s^2, i.e. $I_f^2 t = k^2 s^2$. If we transpose this formula for t, we get $t = k^2 s^2 / I_f^2$, which is the maximum disconnection time in seconds.

Remember that these requirements refer to short-circuit currents only. If in fact the protective device has been selected to protect against overloads and has a breaking capacity not less than the prospective short-circuit current I_p at the point of installation, it will also protect against short-circuit currents. However, if there is any doubt the formula should be used.

For example, in Figure 19, if I_n has been selected for overload protection, the questions to be asked are as follows:

1 Is $I_n \geqslant I_b$? Yes.
2 Is $I_n \leqslant I_z$? Yes.
3 Is $I_2 \leqslant 1.45 I_z$? Yes.

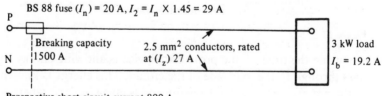

Figure 19

Then if the device has a rated breaking capacity not less than I_p, it can be considered to give protection against short-circuit current also.

When an installation is being designed, the prospective short-circuit current at every relevant point must be determined, by either calculation or measurement. The value will decrease as we move farther away from the intake position (resistance increases with length). Thus if the breaking capacity of the lowest rated fuse in the installation is greater than the prospective short-circuit current at the origin of the supply, there is no need to determine the value except at the origin.

Discrimination
[IEE Reg 533–01–06]

When we discriminate, we indicate our preference over other choices: this house rather than that house, for example. With protection we have to ensure that the correct device operates when there is a fault. Hence a 13 A BS 1362 plug fuse should operate before the main circuit fuse. Logically, protection starts at the origin of an installation with a large device and progresses down the chain with smaller and smaller sizes.

Simply because protective devices have different ratings, it cannot be assumed that discrimination is achieved. This is especially the case where a mixture of different types of device is used. However, as a general rule a 2:1 ratio with the lower-rated devices will be satisfactory. The accompanying table shows how fuse links may be chosen to ensure discrimination.

Fuses will give discrimination if the figure in column 3 does not exceed the figure in column 2. Hence:

a 2A fuse will discriminate with a 4 A fuse
a 4A fuse will discriminate with a 6 A fuse
a 6A fuse will *not* discriminate with a 10 A fuse
a 10A fuse will discriminate with a 16 A fuse

All other fuses will *not* discriminate with the next highest fuse, and in some cases, several sizes higher are needed e.g. a 250 A fuse will only discriminate with a 400 A fuse.

Position of protective devices
[IEE Regs Section 473]

When there is a reduction in the current-carrying capacity of a conductor, a protective device is required. There are, however,

$I_f^2 t$ characteristics: 2–800 A fuse links. Discrimination is achieved if the total $I_f^2 t$ of the minor fuse does not exceed the pre-arcing $I_f^2 t$ of the major fuse

Rating (A)	$I_f^2 t$ pre-arcing	$I_f^2 t$ total at 415 V
2	0.9	1.7
4	4	12
6	16	59
10	56	170
16	190	580
20	310	810
25	630	1 700
32	1 200	2 800
40	2 000	6 000
50	3 600	11 000
63	6 500	14 000
80	13 000	36 000
100	24 000	66 000
125	34 000	120 000
160	80 000	260 000
200	140 000	400 000
250	230 000	560 000
315	360 000	920 000
350	550 000	1 300 000
400	800 000	2 300 000
450	700 000	1 400 000
500	900 000	1 800 000
630	2 200 000	4 500 000
700	2 500 000	5 000 000
800	4 300 000	10 000 000

some exceptions to this requirement; these are listed quite clearly in Section 473 of the IEE Regulations. As an example, protection is not needed in a ceiling rose where the cable size changes from 1.0 mm to, say, 0.5 mm for the lampholder flex. This is permitted as it is not expected that lamps will cause overloads.

Protection against undervoltage
[IEE Regs Chapter 45 and Section 535]

From the point of view of danger in the event of a drop or loss of voltage, the protection should prevent automatic restarting of machinery etc. In fact such protection is an integral part of motor starters in the form of the control circuit.

The essential part of a motor control circuit that will ensure undervoltage protection is the 'hold-on' circuit.

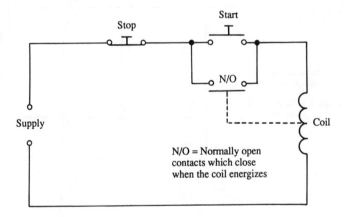

Figure 20 *Hold on circuit*

When the start button is pushed, the coil becomes energized and its normally open (N/O) contacts close. When the start button is released the coil remains energized via its own N/O contacts. These are known as the 'hold-on' contacts.

The coil can only be de-energized by opening the circuit by the use of the stop button or by a considerable reduction or loss of voltage. When this happens, the N/O contacts open and, even if the voltage is restored or the circuit is made complete again, the coil will remain de-energized until the start button is pushed again. Figure 21 shows how this 'hold-on' facility is built into a typical single-phase motor starter.

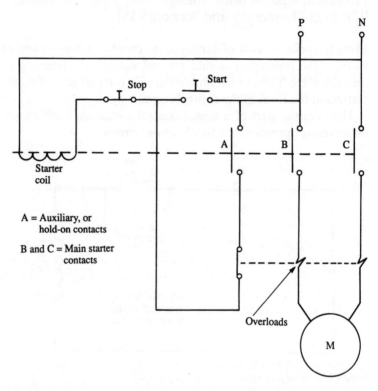

Figure 21 *Single-phase motor starter*

FOUR

Control

Definitions used in this chapter

Emergency switching   Rapid cutting off of electrical energy to remove any hazard to persons, livestock or property which may occur unexpectedly.

Isolation   Cutting off an electrical installation, a circuit, or an item of equipment from every source of electrical energy.

Mechanical maintenance   The replacement, refurbishment or cleaning of lamps and non-electrical parts of equipment, plant and machinery.

Switch   A mechanical switching device capable of making, carrying and breaking current under normal circuit conditions, which may include specified overload conditions, and also of carrying for a specified time currents under specified abnormal conditions such as those of short circuit.

Isolation and switching
[Relevant parts, sections and chapters
Chapter 46. Sections 476 and 537]

Having decided how we are going to earth an installation, and settled on the method of protecting persons and livestock from electric shock, and conductors and insulation from damage, we must now investigate the means of controlling the installation. In simple terms, this means the switching of the installation or any part of it 'on' or 'off'. The IEE Regulations refer to this topic as 'isolation and switching'.

By definition, isolation is the cutting off of electrical energy from every source of supply, and this function is performed by a switch, a switch fuse or a fuse switch. A switch is sometimes referred to as an isolator or a switch disconnector. A switch fuse is a switch housed in the same enclosure as the fuses. A fuse switch has the fuses as part of the switch mechanism, and is used mainly where higher load currents are anticipated.

On the grid system there is a distinct difference between an isolator and a switch. A switch, or breaker as it is known, is capable of breaking fault currents as well as normal load currents. By contrast, an isolator is installed for the purpose of isolating sections of the system for maintenance work etc. (see Figure 22).

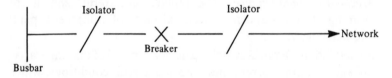

Figure 22

For work to be carried out on the breaker, the part of the network that the circuit supplies is fed from another source. The breaker is then operated and both isolators are opened, thus preventing supply from any source reaching the breaker.

With a domestic installation, the main switch in a consumer unit is considered to be a means of isolation for the whole installation, and each fuse or circuit breaker to be isolators for the individual circuits. Ideally all of these devices should have some means of preventing unintentional re-energization, either by locks or by interlocks. In the case of fuses and circuit breakers, these can be removed and kept in a safe place.

In many cases, isolating and locking off come under the heading of switching off for mechanical maintenance. Hence a switch controlling a motor circuit should have, especially if it is remote from the motor, a means of locking in the 'off' position (Figure 23).

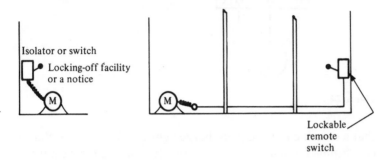

Figure 23

A one-way switch controlling a lighting point is a functional switch, but could be considered as a means of isolation, or a means of switching off for mechanical maintenance (changing a lamp). A two-way switching system, however, does not provide a means of isolation, as neither switch cuts off electrical energy from all sources of supply.

In an industrial or workshop environment it is important to have a means of cutting off the supply to the whole or parts of the installation in the event of an emergency. The most common method is the provision of stop buttons suitably located and used in conjunction with a contactor or relay (Figure 24).

Pulling a plug from a socket to remove a hazard is not permitted as a means of emergency switching. It is however, allowed as a means of functional switching, e.g. switching off a hand lamp by unplugging.

Whilst we are on the subject of switching, it should be noted

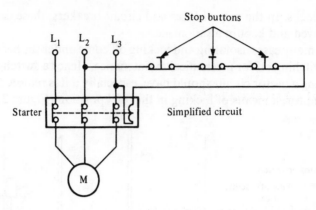

Figure 24

that a switch controlling discharge lighting (this includes fluorescent fittings) should, unless it is specially designed for the purpose, be capable of carrying at least twice the steady load of the circuit. The reason for this is that discharge lighting contains chokes which are highly inductive and cause arcing at switch contacts. The higher rating of the switch enables it to cope with such arcing.

FIVE

Circuit design

Definitions used in this chapter

Ambient temperature   The temperature of the air or other medium where the equipment is to be used.

Circuit protective conductor   A protective conductor connecting exposed conductive parts of equipment to the main earthing terminal.

Current-carrying capacity   The maximum current which can be carried by a conductor under specified conditions without its steady state temperature exceeding a specified value.

Design current   The magnitude of the current intended to be carried by a circuit in normal service.

Earthing conductor   A protective conductor connecting a main earthing terminal of an installation to an earth electrode or other means of earthing.

Overcurrent   A current exceeding the rated value. For conductors the rated value is the current-carrying capacity.

Short-circuit current   An overcurrent resulting from a fault of negligible impedance between live conductors having a difference of potential under normal operating conditions.

Design procedure
[IEE Regs 514–09 and 712–01–03 (xviii); IEE Regs Chapter 3; IEE Regs Appendix 4]

The requirements of IEE Regulations make it clear that circuits must be designed and the design data made readily available. In fact this has always been the case with previous editions of the Regulations, but it has not been so clearly indicated.

How then do we begin to design? Clearly, plunging into calculations of cable size is of little value unless the type of cable and its method of installation is known. This in turn will depend on the installation's environment. At the same time, we would need to know whether the supply was single or three phase, the type of earthing arrangements, and so on. Here then is our starting point, and it is referred to in the Regulations, Chapter 3, as 'Assessment of general characteristics'.

Having ascertained all the necessary details, we can decide on an installation method, the type of cable, and how we will protect against electric shock and overcurrents. We would now be ready to begin the calculation part of the design procedure.

Basically there are eight stages in such a procedure. These are the same whatever the type of installation, be it a cooker circuit or a submain cable feeding a distribution board in a factory. Here then are the eight basic steps in a simplified form:

1 Determine the design current I_b.
2 Select the rating of the protection I_n.
3 Select the relevant correction factors (CFs).
4 Divide I_n by the relevant CFs to give cable current-carrying capacity I_z.
5 Choose a cable size to suit I_z.
6 Check the voltage drop.
7 Check for shock risk constraints.

8 Check for thermal constraints.

Let us now examine each stage in detail.

Design current

In many instances the design current I_b is quoted by the manufacturer, but there are times when it has to be calculated. In that case there are two formulae involved, one for single phase and one for three phase:

Single phase:

$$I_b = \frac{P}{V} \qquad (V \text{ usually } 240\,V)$$

Three phase:

$$I_b = \frac{P}{\sqrt{3} \times V_L} \qquad (V_L \text{ usually } 415\,V)$$

Current is in amperes, and power P in watts.

If an item of equipment has a power factor (PF) and/or has moving parts, efficiency (eff) will have to be taken into account. Hence:

Single phase:

$$I_b = \frac{P \times 100}{V \times PF \times eff}$$

Three phase:

$$I_b = \frac{P \times 100}{\sqrt{3} \times V_L \times PF \times eff}.$$

Nominal setting of protection

Having determined I_b we must now select the nominal setting of the protection I_n such that $I_n \geqslant I_b$. This value may be taken from

IEE Regulations, Tables 41B1, B2 or D or from manufacturers' charts. The choice of fuse or MCB type is also important and may have to be changed if cable sizes or loop impedances are too high. These details will be discussed later.

Correction factors

When a cable carries its full load current it can become warm. This is no problem unless its temperature rises further due to other influences, in which case the insulation could be damaged by overheating. These other influences are: high ambient temperature; cables grouped together closely; uncleared overcurrents; and contact with thermal insulation.

For each of these conditions there is a correction factor (CF) which will respectively be called C_a, C_g, C_f and C_i, and which derates cable current-carrying capacity or conversely increases cable size.

Ambient temperature C_a

The cable ratings in the IEE Regulations are based on an ambient temperature of 30°C, and hence it is only above this temperature that an adverse correction is needed. Table 4C1 of the Regulations gives factors for all types of protection other than BS 3036 semi-enclosed rewirable fuses, which are accounted for in Table 4C2.

Grouping C_g

When cables are grouped together they impart heat to each other. Therefore the more cables there are the more heat they will generate, thus increasing the temperature of each cable. Table 4B of the Regulations gives factors for such groups of cables or circuits. It should be noted that the figures given are for cables of the same size, and hence correction may not necessarily be needed for cables grouped at the outlet of a domestic consumer unit, for example, where there is a mixture of different sizes.

A typical situation where correction factors need to be applied would be in the calculation of cable sizes for a lighting system in a large factory. Here many cables of the same size and loading may be grouped together in trunking and could be expected to be fully loaded all at the same time.

Protection by BS 3036 fuse C_f

As we have already discussed in Chapter 3, because of the high fusing factor of BS 3036 fuses, the rating of the fuse I_n, should be $\leqslant 0.725 I_z$. Hence 0.725 is the correction factor to be used when BS 3036 fuses are used.

Thermal insulation C_i

With the modern trend towards energy saving and the installation of thermal insulation, there may be a need to derate cables to account for heat retention.

The values of cable current carrying capacity given in Appendix 4 of the IEE Regulations have been adjusted for situations when thermal insulation touches one side of a cable. However, if a cable is totally surrounded by thermal insulation for more than 0.5 m, a factor of 0.5 must be applied to the tabulated clipped direct ratings. For less than 0.5 m, derating factors (Table 52A) should be applied.

Application of correction factors

Some or all of the onerous conditions just outlined may affect a cable along its whole length or parts of it, but not all may affect it at the same time. So, consider the following:

1 If the cable in Figure 25 ran for the whole of its length, grouped with others of the same size in a high ambient temperature, and was totally surrounded with thermal insulation, it would seem logical to apply all the CFs, as they all affect the whole cable run. Certainly the factors for the BS 3036 fuse, grouping and thermal insulation should be used. However, it is doubtful if the ambient temperature will have any effect on the cable, as the

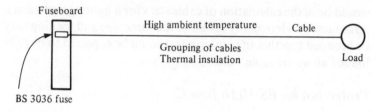

Figure 25

thermal insulation, if it is efficient, will prevent heat reaching the cable. Hence apply C_g, C_f and C_i.

2 In Figure 26a the cable first runs grouped, then leaves the group and runs in high ambient temperature, and finally is enclosed in thermal insulation. We therefore have three different conditions, each affecting the cable in different areas. The BS 3036 fuse affects the whole cable run and therefore C_f must be used, but there is no need to apply all of the remaining factors as the worse one will automatically compensate for the others. The relevant factors are shown in Figure 26b: apply only $C_f = 0.725$

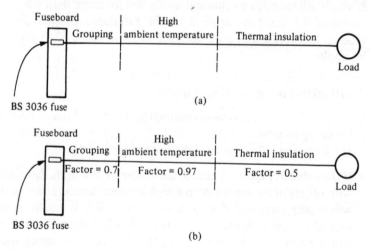

Figure 26

and $C_i = 0.5$. If protection was *not* by BS 3036 fuse, then apply only $C_i = 0.5$.

3 In Figure 27 a combination of cases 1 and 2 is considered. The effect of grouping and ambient temperature is $0.7 \times 0.97 = 0.69$. The factor for thermal insulation is still worse than this combination, and therefore C_i is the only one to be used.

Having chosen the *relevant* correction factors, we now apply them to the nominal rating of the protection I_n as divisors in order to calculate the current-carrying capacity I_z of the cable.

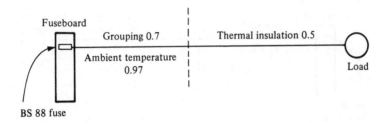

Figure 27

Current-carrying capacity

The required formula for current-carrying capacity I_z is

$$I_z = \frac{I_n}{\text{relevant CFs}}$$

In Figure 28 the current-carrying capacity is given by

$$I_z = \frac{I_n}{C_f C_i} = \frac{30}{0.725 \times 0.5} = 82.75 \text{ A}$$

or, without the BS 3036 fuse,

$$I_z = \frac{30}{0.5} = 60 \text{ A}$$

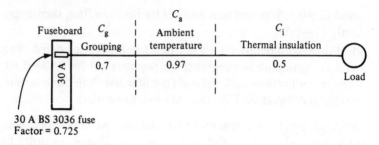

30 A BS 3036 fuse
Factor = 0.725

Figure 28

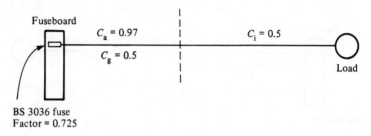

BS 3036 fuse
Factor = 0.725

Figure 29

In Figure 29, $C_a C_g = 0.97 \times 0.5 = 0.485$, which is worse than C_i (0.5). Hence

$$I_z = \frac{I_n}{C_f C_a C_g} = \frac{30}{0.725 \times 0.485} = 85.3 \, \text{A}$$

or, without the B3036 fuse,

$$I_z = \frac{30}{0.485} = 61.85 \, \text{A}$$

Choice of cable size

Having established the current-carrying capacity I_z of the cable to be used, it now remains to choose a cable to suit that value. The tables in Appendix 4 of the IEE Regulations list all the cable sizes, current-carrying capacities and voltage drops of the various types of cable. For example, for PVC-insulated singles, single phase, in conduit, having a current-carrying capacity of 45 A, the installation is by reference method 3 (Table 4A), the cable table is 4D1A and the column is 4. Hence the cable size is $10.0\,mm^2$ (column 1).

Voltage drop
[IEE Regs 525]

The resistance of a conductor increases as the length increases and/or the cross-sectional area decreases. Associated with an increased resistance is a drop in voltage, which means that a load at the end of a long thin cable will not have the full supply voltage available (Figure 30).

The IEE Regulations require that the voltage drop V_c should not be so excessive that equipment does not function safely. They further indicate that a drop of no more than 4 per cent of the nominal voltage at the *origin* of the circuit will satisfy. This means that:

1 For single-phase 240 V, the voltage drop should not exceed 4 per cent of $240 = 9.6\,V$.
2 For three-phase 415 V, the voltage drop should not exceed 4 per cent of $415 = 16.6\,V$.

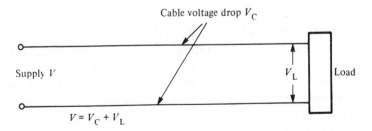

Cable voltage drop V_C

Supply V

V_L Load

$V = V_C + V_L$

Figure 30

For example, the voltage drop on a circuit supplied from a 240 V source by a 16.0 mm two core copper cable 23 m long, clipped direct and carrying a design current of 33 A, will be:

$$V_c = \frac{mV \times I_b \times L}{1000} \quad \text{(mV from Table 4D2B)}$$

$$= \frac{2.8 \times 33 \times 23}{1000} = 2.125 \text{ V}$$

As we know that the maximum voltage drop in this instance (240 V) is 9.6 V, we can determine the maximum length by transposing the formula:

$$\text{maximum length} = \frac{V_c \times 1000}{mV \times I_b}$$

$$= \frac{9.6 \times 1000}{2.8 \times 23} = 149 \text{ m}$$

There are other constraints, however, which may not permit such a length.

Shock risk
[IEE Regs 413–02–04]

This topic has already been discussed in full in Chapter 2. To recap, the actual loop impedance Z_s should not exceed those values given in Tables 41B1, B2 and D of the IEE Regulations. This ensures that circuits feeding socket outlets, bathrooms and equipment outside the equipotential zone will be disconnected, in the event of an earth fault, in less than 0.4 s, and that fixed equipment will be disconnected in less than 5 s.

Remember: $Z_s = Z_e + R_1 + R_2$.

Thermal constraints
[IEE Regs 543]

The IEE Regulations require that we either select or check the size of a CPC against Table 54F, or calculate its size using an adiabatic equation.

Selection of CPC using Table 54G

Table 54G simply tells us that

1 For phase conductors up to and including $16\,\text{mm}^2$, the CPC should be at least the same size.
2 For sizes between $16\,\text{m}^2$ and $35\,\text{mm}^2$, the CPC should be at least $16\,\text{mm}$.
3 For sizes of phase conductor over $35\,\text{mm}^2$, the CPC should be at least half this size.

This is all very well, but for large sizes of phase conductor the CPC is also large and hence costly to supply and install. Also, composite cables such as the typical twin with CPC 6242Y type have CPCs smaller than the phase conductor and hence do not comply with Table 54G.

Calculation of CPC using adiabatic equation

The adiabatic equation

$$s = \sqrt{(I_f^2\, t)}/k$$

enables us to check on a selected size of cable, or on an actual size in a multicore cable. In order to apply the equation we need first to calculate the earth fault current from

$$I_a = U_0/Z_s$$

where U_0 is the nominal voltage to earth (usually 240 V) and Z_s is the actual earth fault loop impedance. Next we select a k factor from Tables 54B to F, and then determine the disconnection time *t* from the relevant curve.

For those unfamiliar with such curves, using them may appear a daunting task. A brief explanation may help to dispel any fears. Let us first refresh our memories with regards to indices such as 10^{-1} and 10^3. These are simply a mathematical way of representing whole numbers, or fractions of a number, as follows:

$10^{-2} =$ 0.01
$10^{-1} =$ 0.1
$10^0 \ =$ 1
$10^1 \ =$ 10
$10^2 \ =$ 100
$10^3 \ =$ 1 000
$10^4 \ =$ 10 000
$10^5 \ =$ 100 000

So, referring to any of the curves in Appendix 3 of the IEE Regulations, we can see that the current scale goes from 1 A (10^0) to 100 000 A (10^5), and the time scale from 0.01 s (10^{-2}) to 10 000 s (10^4). One can imagine the difficulty in drawing a scale between 1 A and 100 000 A in divisions of 1 A, and so a logarithmic scale is used. This cramps the large scale into a small area. All the subdivisions between the major divisions increase in equal amounts depending on the major division boundaries; for example, all the subdivisions between 10^2 (100) and 10^3 (1000) are in amounts of 100 (Figure 31).

Figures 32 and 33 give the IEE Regulations time/current curves for BS 88 fuses. Referring to the appropriate curve for a 32 A fuse (Figure 32), we find that a fault current of 200 A will cause disconnection of the supply in 0.6 s.

Where a value falls between two subdivisions, e.g. 150 A, an estimate of its position must be made. Remember that even if the scale is not visible, it would be cramped at one end; so 150 A would *not* fall half-way between 100 A and 200 A (Figure 34).

It will be noted in Appendix 3 of the Regulations that each set of curves is accompanied by a table which indicates the current that causes operation of the protective device for disconnection times of 0.1 s, 0.4 s and 5 s.

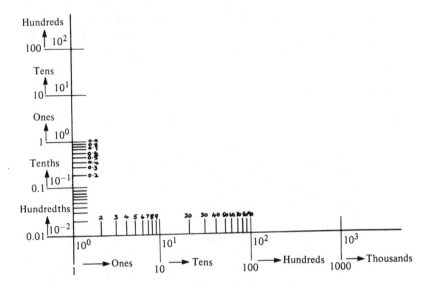

Figure 31

The IEE Regulations curves for MCBs to BS 3871 may give some cause for concern, but they are in fact easily explained (Figures 35, 36 and 37). It will be noted that there are two parts to each curve. This is because MCBs provide protection against overload and short-circuit currents, each of which is performed by a different part of the MCB. An overload is dealt with by a bimetallic mechanism, and a short circuit by an electromagnetic mechanism.

A type 1 MCB can take up to 4 times its rated current on overload before operating instantaneously. The magnetic part of the MCB operates at higher currents. A type 2 MCB can take up to 7 times its rating, a type 3 up to 10 times its rating, and types B and C up to 5 times their ratings.

73

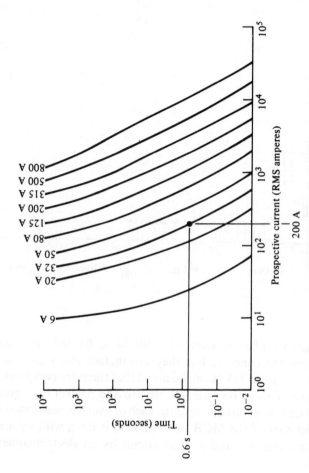

Figure 32 *Time/current characteristics for fuses to BS 88 Part 2. Example for 32 A fuse superimposed*

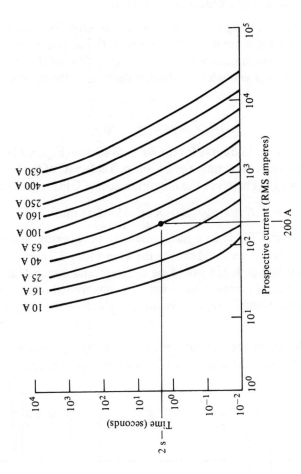

Figure 33 *Time/current characteristics for fuses to BS 88 Part 2. Example for 40 A fuse superimposed*

Having found a disconnection time, we can now apply the formula.

Example of use of adiabatic equation

Suppose that in a design the protection was by 40 A BS 88 fuse; we had chosen a 4 mm copper CPC running with our phase conductor; and the loop impedance Z_s was 1.2 ohms. Would the chosen CPC size be large enough to withstand damage in the event of an earth fault?

We have

$$I_f = U_0/Z_s = 240/1.2 = 200 \text{ A}$$

From the appropriate curve for the 40 A BS 88 fuse (Figure 33), we obtain a disconnection time t of 2 s. From Table 54C of the Regulations, $k = 115$. Therefore the minimum size of CPC is given by

$$s = \sqrt{(I_f^2 t)/k} = \sqrt{(200^2 \times 2)}/115 = 2.46 \text{ mm}^2$$

So our 4 mm^2 CPC is acceptable. Beware of thinking that the answer means that we could change the 4 mm^2 for a 2.5 mm^2. If we did, the loop impedance would be different and hence I_f and t would change; the answer for s would probably tell us to use a 4 mm^2. In the example shown, s is merely a check on the actual size chosen.

Having discussed each component of the design procedure, we can now put all eight together to form a complete design.

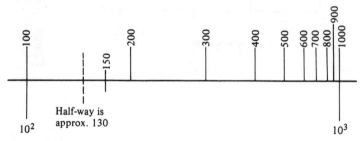

Figure 34

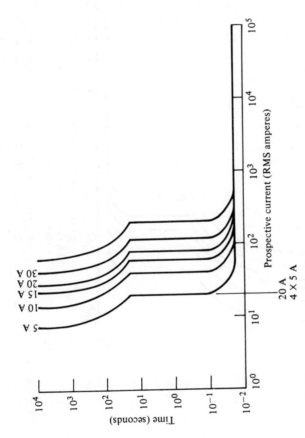

Figure 35 *Time/current characteristics for type 1 MCBs to BS 3871. Example for 20 A superimposed. For times less than 20 ms, the manufacturer should be consulted*

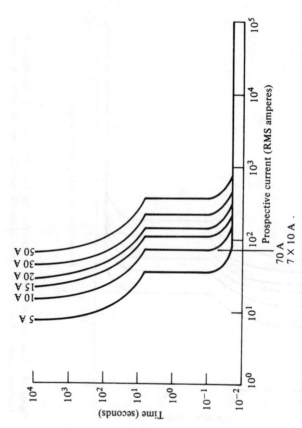

Figure 36 Time/current characteristics for type 2 MCBs to BS3871. Example for 70 A superimposed. For times less than 20 ms, the manufacturer should be consulted

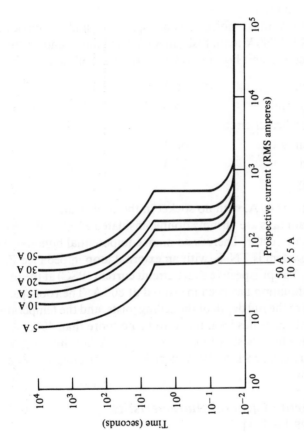

Figure 37 *Time/current characteristics for type 3 MCBs to BS 3871. Example for 50 A superimposed. For times less than 20 ms, the manufacturer should be consulted*

Example of circuit design

A consumer lives in a bungalow with a detached garage and work-shop, as shown in Figure 38. The building method is traditional brick and timber.

The mains intake position is at high level, and comprises an 80 A BS 1361 240 V main fuse, an 80 A rated meter and a six-way 80 A consumer unit housing BS 3036 fuses as follows:

Ring circuit	30 A
Lighting circuit	5 A
Immersion heater circuit	15 A
Cooker circuit	30 A
Shower circuit	30 A
Spare way	

The cooker is 40 A, with no socket in the cooker unit.

The main tails are 16 mm² double-insulated PVC, with a 6 mm earthing conductor. There is no main equipotential bonding. The earthing system is TN–S, with an external loop impedance Z_e of 0.3 ohms. The prospective short-circuit current (PSC) at the origin of the installation has been measured at 800 A. The roof space is insulated to the full depth of the ceiling joists, and the temperature in the roof space has been noted to be no more than 40°C.

The consumer wishes to convert the workshop into a pottery room and install a 9 kW, 240 V electric kiln. The design procedure is as follows.

Assessment of general characteristics
[IEE Regs Part 3]

Present maximum demand
Applying diversity, we have:

Ring	30 A
Lighting (66% of 5 A)	3.3 A
Immersion heater	15 A
A cooker (10 A + 30% of 30 A)	19 A
Shower	30 A
Total	97.3 A

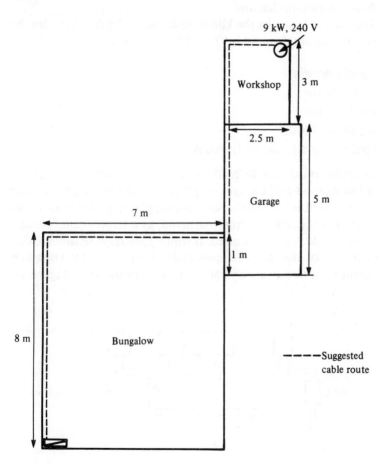

Figure 38

Reference to Table 4D1A of the Regulations will show that the existing main tails are too small and should be uprated. Also, the consumer unit should be capable of carrying the full load of the installation *without* the application of diversity. So the addition of another 9 kW of load is not possible with the present arrangement.

New maximum demand
The current taken by the kiln is $9000/240 = 37.5$ A. Therefore the new maximum demand is $97.3 + 37.5 = 134.8$ A.

Supply details
Single phase
240 V, 50 Hz
Earthing: TN–S
PSC at origin (measured): 800 A

Decisions must now be made as to the type of cable, the installation method and the type of protective device. As the existing arrangement is not satisfactory, the supply authority must be informed of the new maximum demand, as a larger main fuse and service cable may be required. It would then seem sensible to disconnect, say, the shower circuit, and to supply it and the new kiln circuit via a new two-way consumer unit, as shown in Figure 39.

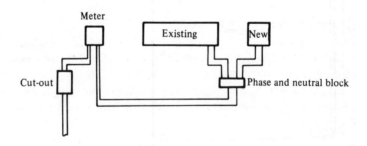

Figure 39

Sizing the main tails
[IEE Reg 547–02]

1 The new load on the existing consumer unit will be the old load less the shower load: $97.3 - 30 = 67.3$ A. From Table 4D1A, the cable size is $16 \, \text{mm}^2$.

2 The load on the new consumer unit will be the kiln load plus the shower load: $37.5 + 30 = 67.5$ A. From Table 4D1A, the cable size is $16 \, \text{mm}^2$.

3 The total load is $67.3 + 67.5 = 134.8$ A. From Table 4D1A, the cable size is $35 \, \text{mm}^2$.

4 The earthing conductor size, from Table 54G, will be $16 \, \text{mm}^2$. The main equipotential bonding conductor size, from Regulation 547–02, will be $10 \, \text{mm}^2$.

For a domestic installation such as this, a PVC flat twin cable clipped direct through the loft space and the garage etc. would be most appropriate.

Sizing the kiln circuit cable

Design current

$$I_b = \frac{P}{V} = \frac{9000}{240} = 37.5 \, \text{A}$$

Rating and type of protection
In order to show how important this choice is, it is probably best to compare the values of current-carrying capacity resulting from each type of protection.

As we have seen, the requirement for the rating I_n is that $I_n \geqslant I_b$. Therefore, using Table 41B2, I_n will be as follows for the various fuse types:

BS 88	40 A	BS 3036	45 A
BS 1361	45 A	BS 3871 MCB	50 A

	BS 88 40 A	BS 1361 45 A	BS 3036 45 A	MCB to BS 3871 50 A
Surrounded by thermal insulation	$\dfrac{40}{0.5 \times 0.87} = 92\,\text{A}$	$\dfrac{45}{0.5 \times 0.87} = 103.4\,\text{A}$	$\dfrac{45}{0.5 \times 0.94 \times 0.725} = 132\,\text{A}$	$\dfrac{50}{0.95 \times 0.87} = 115\,\text{A}$
Not touching	$\dfrac{40}{0.87} = 46\,\text{A}$	$\dfrac{45}{0.87} = 51.7\,\text{A}$	$\dfrac{45}{0.94 \times 0.725} = 66\,\text{A}$	$\dfrac{50}{0.87} = 57.5\,\text{A}$

	BS 88	BS 1361	BS 3036	MCB to BS 3871
Cable size with thermal insulation	$25.0\,\text{mm}^2$	$25.0\,\text{mm}^2$	$35.0\,\text{mm}^2$	$35.0\,\text{mm}^2$
Cable size without	$6.0\,\text{mm}^2$	$10.0\,\text{mm}^2$	$16.0\,\text{mm}^2$	$10.0\,\text{mm}^2$
Cable size with half thermal insulation	$10.0\,\text{mm}^2$	$16.0\,\text{mm}^2$	$25.0\,\text{mm}^2$	$25.0\,\text{mm}^2$

*See item number 15. Table 4A IEE Regulations.
In method 4, correction has already been made for cables touching thermal insulation on one size only.

Correction factors

C_a 0.87 or 0.94 if fuse is BS 3036

C_g not applicable

C_f 0.725 *only* if fuse is BS 3036

C_i 0.5 if cable is totally surrounded in thermal insulation

Current-carrying capacity of cable

For each of the different types of protection, the current-carrying capacity I_z will be as follows:

Cable size based on current-carrying capacity

The table on page 81 shows the sizes of cable for each type of protection (taken from Table 4D2 of the IEE Regulations).

Clearly the BS 88 fuse gives the smallest cable size if the cable is kept clear of thermal insulation.

Check on voltage drop

The actual voltage drop is given by

$$\frac{mV \times I_b \times L}{1000} = \frac{7.3 \times 37.5 \times 24.5}{1000} = 6.7\,V$$

This voltage drop, whilst not causing the kiln to work unsafely, may mean inefficiency, and it is perhaps better to use a $10\,mm^2$ cable. This also gives us a wider choice of protection type, except BS 3036 rewirable. This decision we can leave until later.

For a $10\,mm^2$ cable, the voltage drop is checked as

$$\frac{4.4 \times 37.5 \times 24.5}{1000} = 4.04\,V$$

So at this point we have selected a $10\,mm^2$ twin cable. We have at our disposal a range of protection types, the choice of which will be influenced by the loop impedance.

Shock risk

The CPC inside a $10\,mm^2$ twin 6242Y cable is $4\,mm^2$. Hence the total loop impedance will be

$$Z_s = Z_e + R_1 + R_2$$

For our selected cable, $R_1 + R_2$ for $24.5\,m$ will be (from tables of conductor resistance):

$$\frac{6.44 \times 1.38 \times 24.5}{1000} = 0.218 \text{ ohms}$$

Note: the multiplier 1.38 takes account of the conductor resistance under fault conditions.

We are given that $Z_e = 0.3$ ohms. Hence

$$Z_s = 0.3 + 0.218 = 0.518 \text{ ohms}$$

This means that all but a 50 A type 3 MCB could be used (by comparison of values in Table 41B2 of the Regulations).

Thermal constraints

We still need to check that the $4\,mm^2$ CPC is large enough to withstand damage under earth fault conditions. We have

$$I_a = U_0/Z_s = 240/0.518 = 463 \text{ A}$$

The disconnection time t for each type of protection from the relevant curves in the IEE Regulations are as follows:

BS 88	0.05 s
BS 1361	0.18 s
MCB type 1	0.02 s
MCB type 2	0.035 s

From Table 54C of the Regulations, $k = 115$. Now

$$s = \sqrt{(I_f^2 t)}/k$$

Therefore for each type of protection we have the following sizes s:

BS 88   0.9 mm^2
BS 1361   1.7 mm^2
MCB type 1   0.57 mm^2
MCB type 2   0.75 mm^2

Hence our $4\,\text{mm}^2$ CPC is of adequate size.

Protection

It simply remains to decide on the type of protection. Probably a type 2 MCB is the most economical. However, if this is chosen a check should be made on the shower circuit to ensure that this type of protection is also suitable.

Design problem

In a factory it is required to install, side by side, two three-phase 415 V direct on line motors, each rated at 19 A full load current. There is spare capacity in a three-phase distribution fuseboard housing BS 3036 fuses, and the increased load will not affect the existing installation. The cables are to be PVC-insulated singles installed in steel conduit, and a separate CPC is required (note Regulation 543–01–02). The earthing system is TN–S with a measured external loop impedance of 0.47 ohms, and the length of the cable run is 37 m. The worst conduit section is 7 m long with one bend. The ambient temperature is not expected to exceed 40°C.

Determine the minimum sizes of cable and conduit.

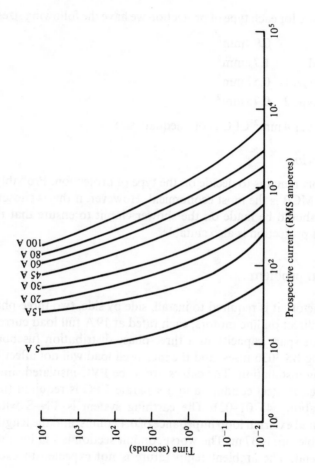

Figure 40 *Time/current characteristics for fuses to BS 1361*

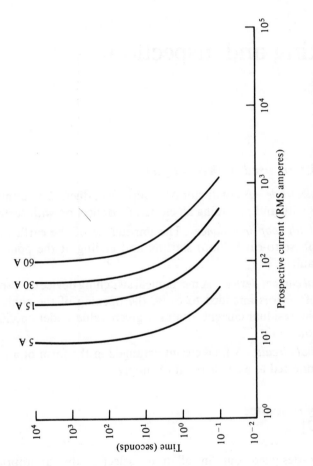

Figure 41 *Time/current characteristics for semi-enclosed fuses to BS 3036*

SIX

Testing and inspection

Definitions used in this chapter

Earth electrode   A conductor or group of conductors in intimate contact with and providing an electrical connection with earth.

Earth fault loop impedance   The impedance of the earth fault loop (phase-to-earth loop) starting and ending at the point of earth fault.

Residual current device   A mechanical switching device or association of devices intended to cause the opening of the contacts when the residual current attains a given value under specified conditions.

Ring final circuit   A final circuit arranged in the form of a ring and connected to a single point of supply.

Testing sequence
[Part 7]

Having designed our installation, selected the appropriate materials and equipment, and installed the system, it now remains to energize it. However, before the installation is energized it must be tested and inspected to ensure that it complies, as far as is practicable, with the IEE Regulations. Note the word 'practicable'; it would be unreasonable, for example, to expect the whole length of a circuit cable to be inspected for defects, as this may mean lifting floorboards etc.

Part 7 of the IEE Regulations give details of testing and inspection requirements. Unfortunately, these requirements pre-suppose that the person carrying out the testing is in possession of all the design data, which is only likely to be the case on the larger commercial or industrial projects. It may be wise for the person who will eventually sign the test certificate to indicate that the test and inspection was carried out as far as was possible in the absence of any design or other information.

However, let us continue by examining the required procedures. The Regulations initially call for a visual inspection, but some items such as correct connection of conductors etc. can be done during the actual testing. A preferred sequence of tests is recommended, where relevant, and is as follows:

1 Continuity of protective conductors.

2 Continuity of ring final circuit conductors.

3 Insulation resistance.

4 Insulation of site-built assemblies.

5 Protection by electrical separation.

6 Protection by barriers and enclosures provided during erection.

7 Insulation of non-conducting floors and walls.

8 Polarity.

9 Earth fault loop impedance.

10 Earth electrode resistance.

11 Operation of residual current devices (RCDs).

One extra test which really should be included is the measurement of the prospective short-circuit current at the origin and/or other relevant points in the installation.

Not all of the tests may be relevant, of course. For example, in a domestic installation (TN–S or TN–C–S) only tests 1, 2, 3, 8, 9 and 11 would be needed.

The Regulations indicate quite clearly in Part 7 the tests required. Let us then take a closer look at some of them in order to understand the reasoning behind them.

Continuity of ring final circuit conductors

Clearly, in order to function correctly, the conductors of a ring final circuit must be continuous. However, a test with an ohm-meter or a bell on the ends of the conductors, whilst indicating continuity, may in fact be incorrect (Figure 42). How then can we test in such a way as to highlight faults of this nature?

The 15th Edition of the Regulations indicated two methods of testing, both of which are based on the principle that resistance changes with length and/or cross-sectional area. Unfortunately, they are also based on the assumption that the position of the socket outlet nearest the midpoint of the ring is known. This should be so for new installations, where this detail is recorded, but not otherwise.

If we do know the position of the midpoint socket, the tests are based on the following principle. Take a conductor and measure

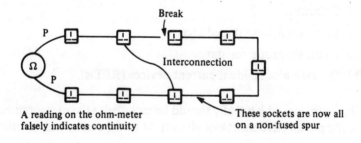

Figure 42

its resistance R_1 (Figure 43). Join A and B together and measure the resistance R_2 between A/B and C (Figure 44). We have now halved the length and doubled the cross-sectional area, and therefore resistance R_2 should be approximately one-quarter of R_1 (half for the reduced length, and half for the increased area). In order to measure such resistances, a very low reading ohmmeter will be required; a digital milliohmmeter is probably best. If R_2 is not approximately one-quarter R_1, this would indicate an interconnection and each socket would have to be investigated to find the fault.

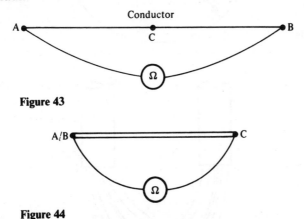

Figure 43

Figure 44

What do we do, however, when the midpoint socket is *not* known? The author has devised a method of testing (now adopted by the IEE and contained in their guidance notes to inspection and testing) that does not require a midpoint socket. It is based on the principle that the resistance measured equidistant across a closed loop will always be the same (Figure 45). The loop is formed by connecting the phase and neutral conductors of *opposite* ends of the ring together (Figure 46). The test procedure is as follows:

1 Check with a milliohmmeter that there is continuity of phase and neutral.

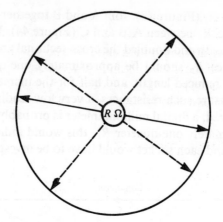

Figure 45

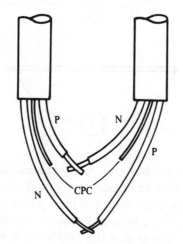

Figure 46

2 Connect the phase and neutral connectors as shown in Figure 46.

3 Check the resistance reading between phase and neutral at every socket on the ring (a suitably adapted plug-top enabling connection of an instrument is useful).

4 If the ring is continuous with no interconnections, all sockets will give the same reading. Spurs give a higher reading.

5 Should an interconnection exist, sockets beyond that interconnection will show a significantly different value. Hence the fault can be located without the systematic removal of all the socket fronts which would be needed with other methods of testing.

Earth electrode resistance

We have already discussed in Chapter 2 the effect of the resistance of an earth electrode. When it forms part of the earth fault loop path its resistance value will need to be measured.

The method used is simply based on the principle of a potential divider (Figure 47). As we move the slider, the voltage will change. Using Ohm's law $R = V/I$, we can determine resistance values.

The same principle is used in electrode resistance measurement. However, here the resistor is replaced by the earth and the slider by various positions of an electrode, and usually the ammeter and voltmeter are replaced by a direct reading ohmmeter.

Insulation resistance

We have seen in Chapter 1 how much more onerous are the requirements of the IEE Regulations in comparison with those of

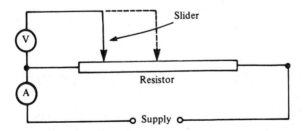

Figure 47

the Electricity Supply Regulations 1988. Most of us will now know that, in general, $0.5\ \text{M}\Omega$ is the minimum value for fixed-wiring and for disconnected equipment (unless specified otherwise by a British Standard).

Remember, insulation resistance is resistance in parallel. Hence, the longer the cable or the greater the number of circuits, the smaller is the insulation resistance. Hence the Regulations allow large installations to be divided up into groups of not less than 50 outlets.

One important point to note is that most modern installations incorporate electronic or microelectronic devices, either in the fixed wiring or in apparatus. Items such as dimmer switches do not take kindly to being subjected to 500 V or 1000 V from an insulation resistance tester; disconnect such items first.

Earth fault loop impedance

As we have seen in Chapter 2, the value of the earth fault loop impedance is of great importance. Its measurement presents no problems, but beware of attempting the test if an RCD is present, as it will see the test as a phase-to-earth fault and of course will trip. Disconnect it first.

Answers to problems

Chapter 2

1 0.292 ohms.
2 18 m.
3 0.5 ohms.
4 No.
5 Yes. Change the cooker control unit to one without a socket, thus converting the circuit to fixed equipment.

Chapter 5

For the factory design problem, the values obtained are as follows:

$I_b = 19$ A; $I_n = 20$ A; $C_f = 0.725$; $C_a = 0.94$; $C_g = 0.8$; $I_z = 36.6$ A; cable size $= 6.0$ mm^2; CPC size $= 2.5$ mm^2; $Z_s = 1$ ohm; $I_f = 240$ A; $t = 1$ s; $k = 115$.

Index